T0192460

LONDON MATHEMATICAL SOCIETY LECTURE NOTE SERIES

Managing Editor: Professor N.J. Hitchin, Mathematical Institute,
University of Oxford, 24–29 St Giles, Oxford OX1 3LB, United Kingdom

The titles below are available from booksellers, or, in case of difficulty, from Cambridge University Press at www.cambridge.org.

113 Lectures on the asymptotic theory of ideals, D. REES
116 Representations of algebras, P.J. WEBB (ed)
119 Triangulated categories in the representation theory of finite-dimensional algebras, D. HAPPEL
121 Proceedings of *Groups - St Andrews 1985*, E. ROBERTSON & C. CAMPBELL (eds)
128 Descriptive set theory and the structure of sets of uniqueness, A.S. KECHRIS & A. LOUVEAU
130 Model theory and modules, M. PREST
131 Algebraic, extremal & metric combinatorics, M.-M. DEZA, P. FRANKL & I.G. ROSENBERG (eds)
138 Analysis at Urbana, II, E. BERKSON, T. PECK, & J. UHL (eds)
139 Advances in homotopy theory, S. SALAMON, B. STEER & W. SUTHERLAND (eds)
140 Geometric aspects of Banach spaces, E.M. PEINADOR & A. RODES (eds)
141 Surveys in combinatorics 1989, J. SIEMONS (ed)
144 Introduction to uniform spaces, I.M. JAMES
146 Cohen-Macaulay modules over Cohen-Macaulay rings, Y. YOSHINO
148 Helices and vector bundles, A.N. RUDAKOV *et al*
149 Solitons, nonlinear evolution equations and inverse scattering, M. ABLOWITZ & P. CLARKSON
150 Geometry of low-dimensional manifolds 1, S. DONALDSON & C.B. THOMAS (eds)
151 Geometry of low-dimensional manifolds 2, S. DONALDSON & C.B. THOMAS (eds)
152 Oligomorphic permutation groups, P. CAMERON
153 L-functions and arithmetic, J. COATES & M.J. TAYLOR (eds)
155 Classification theories of polarized varieties, TAKAO FUJITA
158 Geometry of Banach spaces, P.F.X. MÜLLER & W. SCHACHERMAYER (eds)
159 Groups St Andrews 1989 volume 1, C.M. CAMPBELL & E.F. ROBERTSON (eds)
160 Groups St Andrews 1989 volume 2, C.M. CAMPBELL & E.F. ROBERTSON (eds)
161 Lectures on block theory, BURKHARD KÜLSHAMMER
163 Topics in varieties of group representations, S.M. VOVSI
164 Quasi-symmetric designs, M.S. SHRIKANDE & S.S. SANE
166 Surveys in combinatorics, 1991, A.D. KEEDWELL (ed)
168 Representations of algebras, H. TACHIKAWA & S. BRENNER (eds)
169 Boolean function complexity, M.S. PATERSON (ed)
170 Manifolds with singularities and the Adams-Novikov spectral sequence, B. BOTVINNIK
171 Squares, A.R. RAJWADE
172 Algebraic varieties, GEORGE R. KEMPF
173 Discrete groups and geometry, W.J. HARVEY & C. MACLACHLAN (eds)
174 Lectures on mechanics, J.E. MARSDEN
175 Adams memorial symposium on algebraic topology 1, N. RAY & G. WALKER (eds)
176 Adams memorial symposium on algebraic topology 2, N. RAY & G. WALKER (eds)
177 Applications of categories in computer science, M. FOURMAN, P. JOHNSTONE & A. PITTS (eds)
178 Lower K- and L-theory, A. RANICKI
179 Complex projective geometry, G. ELLINGSRUD *et al*
180 Lectures on ergodic theory and Pesin theory on compact manifolds, M. POLLICOTT
181 Geometric group theory I, G.A. NIBLO & M.A. ROLLER (eds)
182 Geometric group theory II, G.A. NIBLO & M.A. ROLLER (eds)
183 Shintani zeta functions, A. YUKIE
184 Arithmetical functions, W. SCHWARZ & J. SPILKER
185 Representations of solvable groups, O. MANZ & T.R. WOLF
186 Complexity: knots, colourings and counting, D.J.A. WELSH
187 Surveys in combinatorics, 1993, K. WALKER (ed)
188 Local analysis for the odd order theorem, H. BENDER & G. GLAUBERMAN
189 Locally presentable and accessible categories, J. ADAMEK & J. ROSICKY
190 Polynomial invariants of finite groups, D.J. BENSON
191 Finite geometry and combinatorics, F. DE CLERCK *et al*
192 Symplectic geometry, D. SALAMON (ed)
194 Independent random variables and rearrangement invariant spaces, M. BRAVERMAN
195 Arithmetic of blowup algebras, WOLMER VASCONCELOS
196 Microlocal analysis for differential operators, A. GRIGIS & J. SJÖSTRAND
197 Two-dimensional homotopy and combinatorial group theory, C. HOG-ANGELONI *et al*
198 The algebraic characterization of geometric 4-manifolds, J.A. HILLMAN
199 Invariant potential theory in the unit ball of C^n, MANFRED STOLL
200 The Grothendieck theory of dessins d'enfant, L. SCHNEPS (ed)
201 Singularities, JEAN-PAUL BRASSELET (ed)
202 The technique of pseudodifferential operators, H.O. CORDES
203 Hochschild cohomology of von Neumann algebras, A. SINCLAIR & R. SMITH
204 Combinatorial and geometric group theory, A.J. DUNCAN, N.D. GILBERT & J. HOWIE (eds)
205 Ergodic theory and its connections with harmonic analysis, K. PETERSEN & I. SALAMA (eds)
207 Groups of Lie type and their geometries, W.M. KANTOR & L. DI MARTINO (eds)
208 Vector bundles in algebraic geometry, N.J. HITCHIN, P. NEWSTEAD & W.M. OXBURY (eds)
209 Arithmetic of diagonal hypersurfaces over infite fields, F.Q. GOUVÊA & N. YUI
210 Hilbert C^*-modules, E.C. LANCE
211 Groups 93 Galway / St Andrews I, C.M. CAMPBELL *et al* (eds)
212 Groups 93 Galway / St Andrews II, C.M. CAMPBELL *et al* (eds)
214 Generalised Euler-Jacobi inversion formula and asymptotics beyond all orders, V. KOWALENKO *et al*
215 Number theory 1992–93, S. DAVID (ed)

216 Stochastic partial differential equations, A. ETHERIDGE (ed)
217 Quadratic forms with applications to algebraic geometry and topology, A. PFISTER
218 Surveys in combinatorics, 1995, PETER ROWLINSON (ed)
220 Algebraic set theory, A. JOYAL & I. MOERDIJK
221 Harmonic approximation, S.J. GARDINER
222 Advances in linear logic, J.-Y. GIRARD, Y. LAFONT & L. REGNIER (eds)
223 Analytic semigroups and semilinear initial boundary value problems, KAZUAKI TAIRA
224 Computability, enumerability, unsolvability, S.B. COOPER, T.A. SLAMAN & S.S. WAINER (eds)
225 A mathematical introduction to string theory, S. ALBEVERIO, J. JOST, S. PAYCHA, S. SCARLATTI
226 Novikov conjectures, index theorems and rigidity I, S. FERRY, A. RANICKI & J. ROSENBERG (eds)
227 Novikov conjectures, index theorems and rigidity II, S. FERRY, A. RANICKI & J. ROSENBERG (eds)
228 Ergodic theory of Z^d actions, M. POLLICOTT & K. SCHMIDT (eds)
229 Ergodicity for infinite dimensional systems, G. DA PRATO & J. ZABCZYK
230 Prolegomena to a middlebrow arithmetic of curves of genus 2, J.W.S. CASSELS & E.V. FLYNN
231 Semigroup theory and its applications, K.H. HOFMANN & M.W. MISLOVE (eds)
232 The descriptive set theory of Polish group actions, H. BECKER & A.S. KECHRIS
233 Finite fields and applications, S.COHEN & H. NIEDERREITER (eds)
234 Introduction to subfactors, V. JONES & V.S. SUNDER
235 Number theory 1993–94, S. DAVID (ed)
236 The James forest, H. FETTER & B. GAMBOA DE BUEN
237 Sieve methods, exponential sums, and their applications in number theory, G.R.H. GREAVES *et al*
238 Representation theory and algebraic geometry, A. MARTSINKOVSKY & G. TODOROV (eds)
239 Clifford algebras and spinors, P. LOUNESTO
240 Stable groups, FRANK O. WAGNER
241 Surveys in combinatorics, 1997, R.A. BAILEY (ed)
242 Geometric Galois actions I, L. SCHNEPS & P. LOCHAK (eds)
243 Geometric Galois actions II, L. SCHNEPS & P. LOCHAK (eds)
244 Model theory of groups and automorphism groups, D. EVANS (ed)
245 Geometry, combinatorial designs and related structures, J.W.P. HIRSCHFELD *et al*
246 p-Automorphisms of finite p-groups, E.I. KHUKHRO
247 Analytic number theory, Y. MOTOHASHI (ed)
248 Tame topology and o-minimal structures, LOU VAN DEN DRIES
249 The atlas of finite groups: ten years on, ROBERT CURTIS & ROBERT WILSON (eds)
250 Characters and blocks of finite groups, G. NAVARRO
251 Gröbner bases and applications, B. BUCHBERGER & F. WINKLER (eds)
252 Geometry and cohomology in group theory, P. KROPHOLLER, G. NIBLO, R. STÖHR (eds)
253 The q-Schur algebra, S. DONKIN
254 Galois representations in arithmetic algebraic geometry, A.J. SCHOLL & R.L. TAYLOR (eds)
255 Symmetries and integrability of difference equations, P.A. CLARKSON & F.W. NIJHOFF (eds)
256 Aspects of Galois theory, HELMUT VÖLKLEIN *et al*
257 An introduction to noncommutative differential geometry and its physical applications 2ed, J. MADORE
258 Sets and proofs, S.B. COOPER & J. TRUSS (eds)
259 Models and computability, S.B. COOPER & J. TRUSS (eds)
260 Groups St Andrews 1997 in Bath, I, C.M. CAMPBELL *et al*
261 Groups St Andrews 1997 in Bath, II, C.M. CAMPBELL *et al*
263 Singularity theory, BILL BRUCE & DAVID MOND (eds)
264 New trends in algebraic geometry, K. HULEK, F. CATANESE, C. PETERS & M. REID (eds)
265 Elliptic curves in cryptography, I. BLAKE, G. SEROUSSI & N. SMART
267 Surveys in combinatorics, 1999, J.D. LAMB & D.A. PREECE (eds)
268 Spectral asymptotics in the semi-classical limit, M. DIMASSI & J. SJÖSTRAND
269 Ergodic theory and topological dynamics, M.B. BEKKA & M. MAYER
270 Analysis on Lie Groups, N.T. VAROPOULOS & S. MUSTAPHA
271 Singular perturbations of differential operators, S. ALBEVERIO & P. KURASOV
272 Character theory for the odd order function, T. PETERFALVI
273 Spectral theory and geometry, E.B. DAVIES & Y. SAFAROV (eds)
274 The Mandelbrot set, theme and variations, TAN LEI (ed)
275 Computational and geometric aspects of modern algebra, M. D. ATKINSON *et al* (eds)
276 Singularities of plane curves, E. CASAS-ALVERO
277 Descriptive set theory and dynamical systems, M. FOREMAN *et al* (eds)
278 Global attractors in abstract parabolic problems, J.W. CHOLEWA & T. DLOTKO
279 Topics in symbolic dynamics and applications, F. BLANCHARD, A. MAASS & A. NOGUEIRA (eds)
280 Characters and Automorphism Groups of Compact Riemann Surfaces, T. BREUER
281 Explicit birational geometry of 3-folds, ALESSIO CORTI & MILES REID (eds)
282 Auslander-Buchweitz approximations of equivariant modules, M. HASHIMOTO
283 Nonlinear elasticity, R. OGDEN & Y. FU (eds)
284 Foundations of computational mathematics, R. DEVORE, A. ISERLES & E. SULI (eds)
285 Rational points on curves over finite fields: Theory and Applications, H. NIEDERREITER & C. XING
286 Clifford algebras and spinors 2nd edn, P. LOUNESTO
287 Topics on Riemann surfaces and Fuchsian groups, E. BUJALANCE, A. F. COSTA & E. MARTINEZ (eds)
288 Surveys in combinatorics, 2001, J. W. P. HIRSCHFELD (ed)
289 Aspects of Sobolev-type inquealities, L. SALOFF-COSTE
290 Quantum groups and Lie Theory, A. PRESSLEY
291 Tits buildings and the model theory of groups, K. TENT
292 A quantum groups primer, S. MAJID
293 Second order partial differential equations in Hilbert spaces,. . .G. Da Prato & J. Zabczyk

The homotopy category of simply connected 4-manifolds

Hans-Joachim Baues

With an Appendix "On the cohomology of the category **nil**" by Teimuraz Pirashvili.

Shaftesbury Road, Cambridge CB2 8EA, United Kingdom

One Liberty Plaza, 20th Floor, New York, NY 10006, USA

477 Williamstown Road, Port Melbourne, VIC 3207, Australia

314–321, 3rd Floor, Plot 3, Splendor Forum, Jasola District Centre, New Delhi – 110025, India

103 Penang Road, #05–06/07, Visioncrest Commercial, Singapore 238467

Cambridge University Press is part of Cambridge University Press & Assessment, a department of the University of Cambridge.

We share the University's mission to contribute to society through the pursuit of education, learning and research at the highest international levels of excellence.

www.cambridge.org
Information on this title: www.cambridge.org/9780521531030

First published 2003

A catalogue record for this publication is available from the British Library

ISBN 978-0-521-53103-0 Paperback

CONTENTS

INTRODUCTION

We study homotopy classes of maps between simply connected closed topological 4-manifolds. By a well known result of Freedman [F] each simply connected 4-dimensional Poincaré complex X is homotopy equivalent to such a manifold. Here X is a CW-complex

$$X = S^2 \vee \cdots \vee S^2 \cup_g e^4$$

obtained by attaching a 4-cell to a one point union of 2-dimensional spheres. The attaching map g is given by a unimodular form on the free abelian group $A = H_2X$. We call X a (2,4)-Poincaré complex and we consider the homotopy category $\mathbf{P}(2,4)$ consisting of (2,4)-Poincaré complexes and homotopy classes of maps. Then $\mathbf{P}(2,4)$ is equivalent to the homotopy category of simply connected 4-manifolds.

It is an old result that the homotopy type of such a manifold or Poincaré complex X is completely determined by the unimodular form associated to X; compare J.H.C. Whitehead [W4] and Milnor [MI]. The homotopy classes of maps between such Poincaré complexes, however, do not coincide with the homomorphisms between the associated forms. More precisely we obtain for a map $F : Y \to X$ in $\mathbf{P}(2,4)$ the induced maps in homology

$$\begin{aligned} \eta &= F_* : H_2Y \to H_2X, \\ \xi &= F_* : H_4Y = \mathbb{Z} \to H_4X = \mathbb{Z}, \end{aligned}$$

where $\xi \in \mathbb{Z}$ is the degree of F. The pair (ξ, η) is compatible with the unimodular forms f of Y and g of X and we call $(\xi, \eta) : f \to g$ a morphism between unimodular forms. Let **UF** be the category consisting of unimodular forms and such morphisms. Then homology yields a functor

$$H_* : \mathbf{P}(2,4) \to \mathbf{UF}$$

which is full, is representative and reflects isomorphisms. But this functor is not an equivalence of categories.

In fact, let $[Y, X]$ be the set of homotopy classes of maps $Y \to X$ and let

$$[Y, X]_{\xi,\eta} \subset [Y, X]$$

be the subset of all elements $F \in [Y, X]$ which induce $(\xi, \eta) = H_*F$ in homology. If ξ is non-trivial then $[Y, X]_{\xi,\eta}$ is a finite set. We compute an abelian group $D(\xi, \eta)$ which acts transitively and effectively on the set $[Y, X]_{\xi,\eta}$.

For example let

$$Aut(X) \subset [X, X]$$

be the group of homotopy equivalences of X and let $Aut(g)$ be the group of automorphisms of the form g of X in the category **UF**. Then homology H_* yields an extension of groups

$$D(1,1) \rightarrowtail Aut(X) \xrightarrow{H_*} Aut(g).$$

Here H_* is a surjective homomorphism and the kernel of H_* is given by the $(\mathbb{Z}/2)$-vector space

$$D(1,1) \;=\; ker(w : H_2(X, \mathbb{Z}/2) \to \mathbb{Z}/2)$$

defined by the Stiefel-Whitney class w of X. Moreover $D(1,1)$ is an $Aut(g)$-module so that the extension $Aut(X)$ is determined by a cohomology class

$$\{Aut(X)\} \in H^2(Aut(g), D(1,1)).$$

The class $\{Aut(X)\}$ is known to be trivial for all X in $\mathbf{P}(2,4)$ so that there is a homomorphism $Aut(g) \to Aut(X)$ of groups which splits the homomorphism $H_* : Aut(X) \to Aut(g)$.

The extension $Aut(X)$ above, in fact, admits a generalization on the level of categories. The transitive action of $D(\xi,\eta)$ on the set $[Y,X]_{\xi,\eta}$ (denoted by $+$) yields a linear extension of categories

$$D \xrightarrow{+} \mathbf{P}(2,4) \xrightarrow{H_*} \mathbf{UF}$$

which represents a cohomology class

$$\{\mathbf{P}(2,4)\} \in H^2(\mathbf{UF}, D).$$

Here we use the cohomology of categories with coefficients in natural systems introduced in [BW]. The cohomology class determines the extension $\mathbf{P}(2,4)$ up to equivalence. We prove that the cohomolgy class $\{\mathbf{P}(2,4)\}$ is a non trivial element. This implies that there is no functor $s : \mathbf{UF} \to \mathbf{P}(2,4)$ which splits the homology functor $H_* : \mathbf{P}(2,4) \to \mathbf{UF}$. This is a somewhat surprising result since the extension $Aut(X)$ above is known to be split for all X in $\mathbf{P}(2,4)$. Moreover we compute the subextension

$$D \longrightarrow \mathbf{P}(2,4)_0 \longrightarrow \mathbf{UF}_0$$

consisting of morphisms with non trivial degree. We determine the cohomology class $\{\mathbf{P}(2,4)_0\}$ explicitly which is a non trivial element of order 2. For this we describe three algebraic extensions

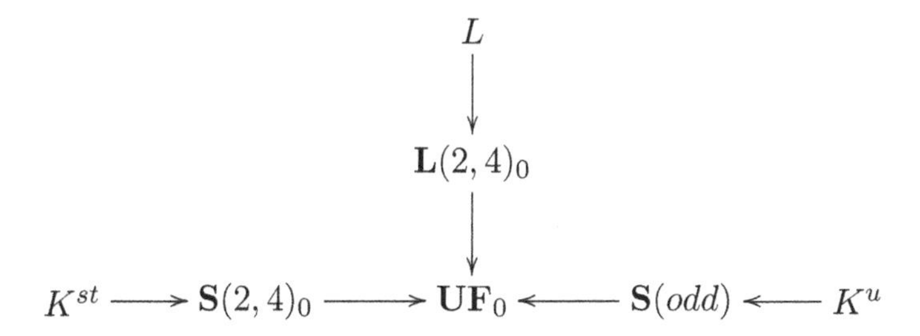

such that $\mathbf{P}(2,4)_0$ is equivalent to the pull back of these three extensions. Compare theorem (2.7.3). This is a complete algebraic characterization of the category $\mathbf{P}(2,4)_0$. As an application this yields the computation of the group $Aut(X)$ of homotopy equivalences of X originally obtained by Cochran and Habegger [CH], see section 2.5.

The category $\mathbf{S}(2,4)_0$ is the *stable part* of $\mathbf{P}(2,4)_0$, that is,

$$\mathbf{S}(2,4)_0 \;=\; \mathbf{P}(2,4)_0/\overset{S}{\simeq}$$

is a quotient category of $\mathbf{P}(2,4)_0$ given by the stable homotopy relation $\overset{S}{\simeq}$. Here we have $F \overset{S}{\simeq} G$ for $F, G : Y \to X$ if the double suspensions $\Sigma^2 F, \Sigma^2 G$ are homotopic.

Next the category $\mathbf{L}(2,4)$ is the *Lie part* of $\mathbf{P}(2,4)_0$. The computation of the Lie part heavily relies on the methods in the book [BCU]. Let **nil** be the subcategory of groups consisting of groups $F/\Gamma_3 F$ where F is a finitely generated free group and $\Gamma_3 F$ is the subgroup of triple commutators. Then $\mathbf{L}(2,4)_0$ is algebraically derived from the category **nil**.

Finally the *odd part* $\mathbf{S}(odd)$ of $\mathbf{P}(2,4)_0$ is determined by the "signature derivation". The computation of $\mathbf{S}(odd)$ relies on a result of Rochlin [R] concerning the role of $\mathbb{Z}/16$ for simply connected 4-manifolds.

We also study the homotopy category $\mathbf{CW}(2,4)$ of (2,4)-complexes which are CW-complexes X of the form

$$X = S^2 \vee \cdots \vee S^2 \cup e^4 \cup \cdots \cup e^4$$

obtained by attaching a collection of 4-cells to a one-point union of 2-spheres. Of course (2,4)-Poincaré complexes are special (2,4)-complexes so that we have the inclusion of categories

$$\mathbf{P}(2,4) \subset \mathbf{CW}(2,4).$$

In chapter 1 we show that $\mathbf{CW}(2,4)$ is also part of a linear extension of categories

$$D \longrightarrow \mathbf{CW}(2,4) \xrightarrow{H_*} \mathbf{H}(2,4)$$

where H_* is the homology functor.

As a further main result in this book we compute an algebraic category $\mathbf{T}$ together with a natural equivalence relation $\simeq$ such that the quotient category $\mathbf{T}/\simeq$ admits an equivalence

$$\mathbf{T}/\simeq \; = \; \mathbf{CW}(2,4)$$

of categories which is also an equivalence of linear extensions. Compare chapter 5. The algebraic model $\mathbf{T}/\simeq$ of the category $\mathbf{CW}(2,4)$ relies on the construction of certain quadratic refinements of J.H.C. Whitehead's quadratic functor Γ in [W].

The computation of the algebraic category $\mathbf{T}/\simeq$, equivalent to $\mathbf{CW}(2,4)$, is an application of the theory on crossed modules and quadratic modules in the book [BCU]. The category $\mathbf{T}/\simeq$ yields a sophisticated algebraic description of the categories

$$\mathbf{P}(2,4)_0 \subset \mathbf{CW}(2,4) = \mathbf{T}/\simeq .$$

In fact, this is needed in order to compute the Lie part $\mathbf{L}(2,4)_0$ of $\mathbf{P}(2,4)_0$ above. We, however, do not achieve our result on the odd part $\mathbf{S}(odd)$ of $\mathbf{P}(2,4)_0$ by use of the category $\mathbf{T}$. Such a computation (independent of Rochlin's result [R]) remains an open question.

There has been great interest in the literature in constructing algebraic model categories of homotopy theory, in particular of rational homotopy theory. Over the integers $\mathbb{Z}$, however, there are only a very few explicit computations of homotopy categories in the literature. Our computation of $\mathbf{P}(2,4)_0$ and $\mathbf{CW}(2,4)$ achieves such algebraic characterization of homotopy categories over $\mathbb{Z}$.

Bonn, January 2001

Hans-Joachim Baues

CHAPTER 1

THE HOMOTOPY CATEGORY OF (2,4)-COMPLEXES

A (2,4)-complex is a CW-complex X with cells only in dimension 2 and 4. Hence X is of the form

$$X = S^2 \vee \cdots \vee S^2 \cup e^4 \cup \cdots \cup e^4$$

where we attach 4-cells to a one point union of 2-spheres. It is well known that a simply connected 4-manifold M is homotopy equivalent to a (2,4)-complex with only one 4-cell. Moreover if N is a simply connected 6-manifold with torsion free homology and $H_3 N = 0$ then the 4-skeleton of N is a (2,4)-complex. For example the 4-skeleton of the product $N = S^2 \times S^2 \times S^2$ is the (2,4)-complex

$$N^4 = S^2 \times S^2 \times * \cup S^2 \times * \times S^2 \cup * \times S^2 \times S^2.$$

In this chapter we consider the homotopy category $\mathbf{CW}(2,4)$ of (2,4)-complexes and we show that $\mathbf{CW}(2,4)$ is part of a linear extension of categories. In chapter 5 we will describe an algebraic category which is equivalent to $\mathbf{CW}(2,4)$.

1.1. Quadratic functions and the Hopf map

We recall from the literature some basic facts on low dimensional homotopy groups π_3 and π_4. J.H.C. Whitehead [W] showed that the Hopf map $\eta : S^3 \to S^2$ has a quadratic distributivity law. Therefore the quadratic functor Γ can be used to describe the homotopy groups π_3 and π_4 of a Moore space $M(A, 2)$ where A is a free abelian group.

A function $f : A \to B$ between abelian groups is called *quadratic* if $f(-a) = f(a)$ and if the function $A \times A \to B$, given by $(a,b) \mapsto f(a+b) - f(a) - f(b)$, is bilinear. There is the universal quadratic map

$$\gamma : A \longrightarrow \Gamma(A) \tag{1.1.1}$$

with the property: for all quadratic maps $f : A \to B$ there is a unique (induced) homomorphism $f^{\square} : \Gamma(A) \to B$ with $f^{\square}\gamma = f$. A homomorphism $\varphi : A' \to A$ yields the quadratic map $\gamma\varphi$ which induces $\Gamma(\varphi) = (\gamma\varphi)^{\square} : \Gamma(A') \to \Gamma(A)$.

This shows that $\Gamma : \mathbf{Ab} \to \mathbf{Ab}$ is a well defined functor, where $\mathbf{Ab}$ denotes the category of abelian groups. We associate with Γ the following natural commutative diagram in which the subdiagram "push" is a push out of abelian groups; moreover $\otimes^2 A = A \otimes A$ denotes the *tensor product* over $\mathbb{Z}$.

(1.1.2)

$$\begin{array}{ccccc}
A \otimes A & & & & \\
\downarrow{\scriptstyle [1,1]} & & & & \\
\Gamma(A) & \xrightarrow{\tau} & A \otimes A & \xrightarrow{q} & A \wedge A \\
{\scriptstyle \sigma}\downarrow & & \downarrow{\scriptstyle \bar{\sigma}} & & \downarrow{\scriptstyle id} \\
A \otimes \mathbb{Z}/2 & \xrightarrow{\bar{\tau}} & A \hat{\otimes} A & \xrightarrow{q} & A \wedge A
\end{array}$$

The column and the rows of the diagram are exact sequences of abelian groups. The homomorphism τ is induced by the quadratic map $A \to A \otimes A$, $A \mapsto a \otimes a$, and σ is induced by $a \mapsto a \otimes 1$. Here $1 \in \mathbb{Z}/2$ ist the generator of $\mathbb{Z}/2 = \mathbb{Z}/2\mathbb{Z}$. We define $[1,1]$ by

$$[1,1](a \otimes b) = [a,b] = \gamma(a+b) - \gamma(a) - \gamma(b). \tag{1}$$

We clearly have $[a,b] = [b,a]$, $[a,a] = 2\gamma(a)$, $\Gamma(\varphi)[a,b] = [\varphi a, \varphi b]$, $\sigma[a,b] = 0$ and $\tau[a,b] = a \otimes b + b \otimes a$.

We obtain the *exterior product* $\Lambda^2 A = A \wedge A = A \otimes A/\sim$ by the relation $\tau\gamma(a) = a \otimes a \sim 0$ and we obtain $\hat{\otimes}^2 A = A \hat{\otimes} A = A \otimes A/\sim$ by the relation $\tau[a,b] = a \otimes b + b \otimes a \sim 0$. The quotient map q carries $a \otimes b$ to $a \wedge b$.

If A is free abelian with ordered basis Z, then ΓA and $A \wedge A$ are free abelian with the basis $\{\gamma(m), [m,n] : m < n,\ m,n \in Z\}$ and $\{m \wedge n : m < n,\ m,n \in Z\}$ respectively. In this case the homomorphisms τ ans $\bar{\tau}$ in (1.1.2) are injective. If A is free abelian we define the (integral) *matrix* $(x_{m,n} : m,n \in Z)$ *of an element* $x \in \Gamma A$ by the formula

$$\text{(2)} \qquad \tau(x) = \sum_{m,n \in Z} x_{m,n} m \otimes n.$$

We say that x is *unimodular* if Z is a finite set and if the matrix of x satisfies determinant$(x_{m,n}) \in \{1,-1\}$. The matrix $(x_{m,n})$ of x is a symmetric integral matrix since $x_{m,n} = x_{n,m}$ which determines the element x by

$$x = (\sum_{m<n} x_{m,n}[m,n]) + (\sum_{m} x_{m,m}\gamma(m)).$$

This way we identify symmetric integral matrices with elements in $\Gamma(A)$.

Recall that a *Moore space* $M(A,n)$, $n \geq 2$, is a simply connected CW-complex with reduced homology groups $H_i M(A,n) = 0$ for $i \neq n$ and $H_n M(A,n) = A$. For a free abelian group A with basis Z we set

$$\text{(1.1.3)} \qquad M(A,n) = \bigvee_{Z} S^n$$

where the right hand side is a one point union of n-spheres over the index set Z. The element $m \in Z$ yields the inclusion $m : S^n \subset M(A,n)$ also denoted by m. This is compatible with the natural identification $A = H_n M(A,n) = \pi_n M(A,n)$. Recall that for pointed spaces X, Y we have the *set* $[X,Y]$ *of homotopy classes* of basepoint preserving maps $X \to Y$, for example $[S^n, Y] = \pi_n Y$ is the nth homotopy group of Y. For a free abelian group A we identify

$$\text{(1)} \qquad [M(A,n), Y] = Hom(A, \pi_n Y),$$

so that

$$\text{(2)} \qquad [M(A,n), M(B,n)] = Hom(A,B)$$

Here $Hom(A,B)$ is the abelian group of homomorphisms from A to B. The isomorphism (1) carries a map f to the induced homomorphism $\pi_n(f)$.

The *Hopf map* $\eta : S^3 \to S^2$ induces a quadratric map $\eta^* : \pi_2 Y \to \pi_3 Y$ satisfying the formula

$$\eta^*(a+b) = \eta^*(a) + \eta^*(b) + [a,b] \tag{1.1.4}$$

where $[a,b]$ is the Whitehead product. The quadratic map η^* induces a homomorphism

$$i = (\eta^*)^{\square} : \Gamma(\pi_2 Y) \to \pi_3 Y$$

between abelian groups. This homomorphism is part of Whitehead's certain exact sequence below.

Let Y be a connected CW-complex with skeletons Y^n and basepoint $* \in Y^0$ and let $\hat{Y}$ be the universal covering of Y. Then the Hurewicz homomorphism $(n \geq 2)$

$$h = h_Y : \pi_n Y \cong \pi_n \hat{Y} \to H_n \hat{Y}$$

is part of the long exact sequence

$$H_{n+1}\hat{Y} \xrightarrow{b} \Gamma_n Y \xrightarrow{i} \pi_n Y \xrightarrow{h} H_n \hat{Y} \xrightarrow{b} \Gamma_{n-1} Y$$

where the groups $\Gamma_n Y$ are defined by $\Gamma_n Y = im(\ \pi_n Y^{n-1} \xrightarrow{i_*} \pi_n Y^n\ )$ with i_* induced by the inclusion $i : Y^{n-1} \hookrightarrow Y^n$. One has a quadratic map $\eta^* : \pi_2 Y \to \Gamma_3 Y$ which carries $\alpha : S^2 \to Y^2 \subset Y^3$ to $\alpha\eta$ and the induced map

$$\Gamma(\pi_2 Y) \xrightarrow{\cong} \Gamma_3 Y$$

is an isomorphism for all connected spaces Y. Hence one has the exact sequence

$$\begin{array}{l} H_5\hat{Y} \xrightarrow{b} \Gamma_4 Y \xrightarrow{i} \pi_4 Y \xrightarrow{h} H_4 \hat{Y} \xrightarrow{b} \\ \Gamma(\pi_2 Y) \xrightarrow{i} \pi_3 Y \xrightarrow{h} H_3 \hat{Y} \longrightarrow 0. \end{array} \tag{1.1.5}$$

The operator $b = b_Y$ is the secondary boundary operator of Whitehead, see

[W], [BAH], [BCH], [BHH]. As an application of (1.1.5) one gets the natural isomorphism

$$(1.1.6) \qquad i : \Gamma(A) = \pi_3 M(A,2)$$

which we use as an identification. This way the map γ in (1.1.1) corresponds to η^* and [1, 1] in (1.1.2) corresponds to the Whitehead product $\pi_2 \otimes \pi_2 \to \pi_3$. Moreover σ in (1.1.2) corresponds to the suspension homomorphism Σ,

$$(1.1.7) \qquad \sigma : \Gamma(A) = \pi_3 M(A,2) \xrightarrow{\Sigma} \pi_4 M(A,3) = A \otimes \mathbb{Z}/2.$$

We now consider the next homotopy group $\pi_4 M(A,2)$. For this we need the following definition.

1.1.8. Definition: Let $T(A,1)$ be the free graded tensor algebra generated by the abelian group A where A is concentated in degree 1, that is $T(A,1)_n = \otimes^n A$ is the n-fold tensor product of A. We define the structure of a graded Lie algebra on $T(A,1)$ by

$$(1) \qquad [x,y] \;=\; xy - (-1)^{|x||y|} yx$$

for $x, y \in T(A,1)$. As usual we set $|x| = n$ if $x \in T(A,1)_n$. Let $L(A,1)$ be the sub Lie algebra generated by A in $T(A,1)$ and let $L(A,1)_n = L(A,1) \cap \otimes^n A$. Clearly $L(\_,1)_n$ is a functor **Ab** $\to$ **Ab**. We obtain

$$(2) \qquad L(A,1)_3 \;=\; image[[1,1],1] : \otimes^3 A \to \otimes^3 A$$

where $[[1,1],1]$ carries $a \otimes b \otimes c$ to $[[a,b],c] = (a \otimes b + b \otimes a) \otimes c - c \otimes (a \otimes b + b \otimes a)$. Hence we have the quotient map

$$(3) \qquad \otimes^3 A \twoheadrightarrow L(A,1)_3$$

which carries $a \otimes b \otimes c$ to $[[a,b],c]$ The kernel of this map is generated by the following elements:

(a) $a \otimes b \otimes c + c \otimes a \otimes b + b \otimes c \otimes a$,
(b) $a \otimes b \otimes c - b \otimes a \otimes c$,
(c) $a \otimes a \otimes a$.

Compare 11.1.5 in [BHH]

1.1.9. PROPOSITION: *Let A be a free abelian group. Then one has natural isomorphisms*

$$\begin{aligned}\pi_4 M(A,2) \cong \Gamma_2^2 A &= \Gamma(A)\otimes\mathbb{Z}/2 \oplus L(A,1)_3\\ &\cong (\Gamma(A)\otimes\mathbb{Z}/2\oplus\Gamma(A)\otimes A)/R(A).\end{aligned}$$

Here the natural subgroup $R(A)$ is generated by the following elements ($x, y, z \in A$).

(i) $[x,y]\otimes z \;+\; [z,x]\otimes y \;+\; [y,z]\otimes x$,
(ii) $(\gamma x)\otimes x$,
(iii) $[x,y]\otimes 1 \;+\; (\gamma x)\otimes y \;+\; [y,x]\otimes x$.

Proof: For $\pi_i = \pi_i M(A,2)$ we have the natural homomorphisms

(1) $$(\Sigma\eta)^* : \Gamma(A)\otimes\mathbb{Z}/2 = \pi_3\otimes\mathbb{Z}/2 \longrightarrow \pi_4,$$

(2) $$w : \Gamma(A)\otimes A = \pi_3\otimes\pi_2 \longrightarrow \pi_4$$

which are defined by $\Sigma\eta : S^4 \to S^3$ and by the Whitehead product $w = [-,-]$ respectively. Moreover we define

(3) $$[[1,1],1] : \otimes^3 A \xrightarrow{w'} \Gamma(A)\otimes A \longrightarrow \pi_4$$

by the triple Whitehead product, that is $w' = [1,1]\otimes 1$. This map induces the inclusion $[[1,1],1] : L(A,1)_3 \subset \pi_4$ which together with (1) yields the first isomorphism in (1.1.9). The second isomorphism is induced by (1) and (2); this shows that (i) corresponds to the Jacobi identity for Whitehead products, (ii) corresponds to the identity $[\eta, i_2] = 0$ for generators $\eta\in\pi_3(S^2)$, $i_2\in\pi_2(S^2)$ and (iii) corresponds to a special case of the Barcus-Barratt formula

(4) $$[x\eta, y] = [x,y]\Sigma\eta - [[y,x],x].$$

Using the Hilton Milnor theorem one can check that the maps in (1.1.9) are actually isomorphisms, compare [BCC]. The splitting for $L(A,1)_3$ is obtained by the James-Hopf invariant γ_3 for which the composition

$$(5) \qquad L(A,1)_3 \xrightarrow{[[1,1],1]} \pi_4 M(A,2) \xrightarrow{\gamma_3} \otimes^3 A$$

is the inclusion in (1.1.8)(2). Compare also 11.1.9 of [BHH] and [BG]. □

We now consider the double suspension

$$(1.1.10) \qquad \Sigma^2 : \pi_4 M(A,2) \longrightarrow \pi_6 M(A,4) = A \otimes \mathbb{Z}/2$$

which is trivial on $L(A,1)_3$ and $\Gamma A \otimes A$ and which is given by $\sigma \otimes \mathbb{Z}/2 : \Gamma(A) \otimes \mathbb{Z}/2 \to A \otimes \mathbb{Z}/2 \otimes \mathbb{Z}/2$ on $\Gamma(A) \otimes \mathbb{Z}/2$, see (1.1.7). As usual we use the notation $f \otimes A = f \otimes 1_A$ where $1_A = 1$ is the identity of A and where f is a homomorphism.

1.2. Simply connected 4-manifolds and (2, 4)-complexes

We say that a CW-complex X is a (2,4)-*complex* if $X^0 = *$ and if $X - *$ consists only of 2-cells and 4-cells. Then X is the mapping cone $X = C_g$ of an attaching map

$$(1.2.1) \qquad g : M(B,3) \longrightarrow M(A,2)$$

where $A = H_2 X$ and $B = H_4 X$ are free abelian groups, see (1.1.3). The homotopy class of g is a homomorphism

$$(1) \qquad b_X = g : B = H_4 X \longrightarrow \Gamma(A) = \Gamma(H_2 X)$$

via (1.1.3)(1) and (1.1.6). Here g coincides with the secondary boundary b_X in Whitehead's exact sequence (1.1.5). The *cup product* pairing $\cup$ for the cohomology $H^*(X)$ is obtained by the commutative diagram

$$(2) \qquad \begin{array}{ccc} H^2(X) \otimes H^2(X) & = & Hom(A,\mathbb{Z}) \otimes Hom(A,\mathbb{Z}) \\ \cup \downarrow & & \downarrow (\tau g)^* \\ H^4(X) & = & Hom(B,\mathbb{Z}) \end{array}$$

where $(\tau g)^*(\alpha \otimes \beta) = (\alpha \otimes \beta)\tau g$, see (1.1.2). This shows that the homotopy class of g in (1.2.1) is also determined by the cohomology ring $H^*(X)$.

We say that a (2,4)-complex X as above is a (2,4)-*Poincaré complex* if $H_4X = B = \mathbb{Z}$, with the *fundamental class* $[X] \in H_4X$ as a generator, and if the element

(3) $$b_X[X] \;=\; g(1) \in \Gamma(A) = \Gamma(H_2X)$$

is unimodular, see (1.1.2)(2). Here we assume A to be a finitely generated free abelian group with basis Z. We call $b_X[X]$ the *quadratic form* of X.

Let Z be a basis of A and Z^* be the dual basis of $H^2X = Hom(A, \mathbb{Z})$. Then the matrix $(g_{m,n})$ of $g(1) \in \Gamma(A)$ with respect to Z is also the matrix of the cup product pairing with respect to the dual basis Z^*, that is for $m^* \in Z^*$ corresponding to $m \in Z$ one gets

(4) $$(m^* \cup n^*)([X]) \;=\; g_{m,n}.$$

We have the *Poincaré duality isomorphism*

(5) $$\psi = \psi_g : H_2X \cong H^2X$$

which carries $z \in H_2X = A$ to the unique element $x \in H^2X = Hom(A, \mathbb{Z})$ with $(x \otimes A)\tau g(1) = z$. Here we use the composite

$$\Gamma(A) \xrightarrow{\;\tau\;} A \otimes A \xrightarrow{\;x \otimes A\;} \mathbb{Z} \otimes A = A.\cdot$$

The *intersection form* of X is the bilinear form

(6) $$\cap : H_2X \otimes H_2X \to \mathbb{Z}$$

which carries $a \otimes b$ to $(\psi a) \cup (\psi b)$. Then $\psi^{-1}Z^*$ is the *Poincaré dual basis* of Z in $A = H_2X$. The matrix of the intersecion form with respect to $\psi^{-1}Z^*$ is $(g_{m,n})$ by (4).

We consider the full inclusion of homotopy categories

(1.2.2) $$\mathbf{P}(2,4) \subset \mathbf{CW}(2,4) \subset \mathbf{Top}^*/\simeq .$$

Here $\mathbf{Top}^*$ is the category of pointed topological spaces; the morphisms $X \to Y$ in the homotopy category $\mathbf{Top}^*/\simeq$ are the elements of the homotopy set $[X, Y]$. The categories $\mathbf{CW}(2,4)$ and $\mathbf{P}(2,4)$ are the full subcategories of $\mathbf{Top}^*/\simeq$ consisting of (2,4)-complexes and of (2,4)-Poincaré complexes respectively. By a result of Freedman [F] any (2,4)-Poincaré complex has the homotopy type of a 1-*connected* 4-*dimensional closed topological manifold* so that $\mathbf{P}(2,4)$ is equivalent to the homotopy category of such manifolds.

We now compare the maps in $\mathbf{CW}(2,4)$ with the induced maps in homology. For this we introduce the category $\mathbf{H}(2,4)$. The objects of this category are homomorphisms $g : B \to \Gamma(A)$ where A and B are free abelian groups. The morphisms $(\xi, \eta) : f \to g$ are given by commutative diagrams

$$\begin{array}{ccc} B' & \xrightarrow{\xi} & B \\ {\scriptstyle f}\downarrow & & \downarrow{\scriptstyle g} \\ \Gamma(A') & \xrightarrow[\Gamma(\eta)]{} & \Gamma(A) \end{array} \tag{1}$$

There is the *homology functor*

$$H_* : \mathbf{CW}(2,4) \to \mathbf{H}(2,4) \tag{2}$$

which carries a $(2,4)$-complex X to the homotopy class $g = b_X$ of its attaching map, that is

$$H_*(X) \;=\; \{b_X : H_4X \to \Gamma(H_2X)\} \tag{3}$$

is given by Whitehead's exact sequence. The naturality of this sequence shows that H_* is a well defined functor. The restriction of this functor to (2,4)-Poincaré complexes yields the functor

$$H_* : \; \mathbf{P}(2,4) \longrightarrow \mathbf{UF} \tag{4}$$

where $\mathbf{UF}$ is the full *subcategory of unimodular forms* in $\mathbf{H}(2,4)$. The objects of $\mathbf{UF}$ are homomorphisms $g : B = \mathbb{Z} \to \Gamma(A)$ for which $g(1)$ is unimodular, see (1.2.1)(3). A morphism $(\xi, \eta) : f \to g$ in $\mathbf{UF}$ yields an integer $\xi \in \mathbb{Z}$ which we call the *degree*. For a map $F : Y \to X$ between (2,4)-Poincaré complexes with $H_*(F) = (\xi, \eta)$ we have $(H_4F)[Y] = \xi[X]$ and $H_2F = \eta$. We point out that for a morphism (ξ, η) in **UF** the map η determines ξ. In fact, in the commutative diagram

$$\begin{array}{ccc} \mathbb{Z} & \xrightarrow{\xi} & \mathbb{Z} \\ f\downarrow & & \downarrow g \\ \Gamma(A') & \xrightarrow[\Gamma(\eta)]{} & \Gamma(A) \end{array}$$

the maps f and g are injective so that there is a unique element $\xi \in \mathbb{Z}$ with

$$\Gamma(\eta)(f(1)) = \xi \cdot g(1). \tag{5}$$

Recall that a functor $F : \mathbf{C} \to \mathbf{K}$ is *full* if F is surjective on morphism sets. Moreover F is *faithful* if F is injective on morphism sets. The functor F is *representative* if for each object K in $\mathbf{K}$ there exists an object X in $\mathbf{C}$ together with an isomorphism $F(X) \cong K$ in $\mathbf{K}$. The functor F *reflects isomorphisms* if a map $f : X \to Y$ in $\mathbf{C}$ is an isomorphism in $\mathbf{C}$ if and only if the induced map $F(f) : F(X) \to F(Y)$ is an isomorphism in $\mathbf{K}$.

1.2.3. PROPOSITION: *The homology functor*

$$H_* : \mathbf{CW}(2,4) \to \mathbf{H}(2,4)$$

is full and representative and H_* reflects isomorphisms. The functor H_* is not faithfull.

The proposition implies that H_* induces a bijection between homotopy types of (2,4)-complexes and isomorphism classes of objects in **H**(2,4).

We now consider the homotopy groups π_3, π_4 which are functors

$$\pi_3, \pi_4 : \mathbf{CW}(2,4) \to \mathbf{Ab} \tag{1.2.4}$$

For a (2,4)-complex X these groups are part of the exact sequence

$$0 \longrightarrow \Gamma_4 X \longrightarrow \pi_4 X \longrightarrow H_4 X \xrightarrow{g} \Gamma(H_2 X) \longrightarrow \pi_3 X \longrightarrow 0.$$

Hence we get the natural isomorphism

$$\pi_3 X \;=\; cok(g) \tag{1}$$

where $g = b_X$ is the attaching map of X. Moreover one has a non-natural isomorphism

$$\pi_4 X \;\cong\; \Gamma_4(X) \oplus kernel(g) \tag{2}$$

since $kernel(g)$ is a free abelian group. Here the group $\Gamma_4(X) = \Gamma_2^2(g)$ can be described naturally in terms of g as follows

1.2.5. Definition: For $g : B \to \Gamma(A)$ in $\mathbf{H}(2{,}4)$ we define $\Gamma_2^2(g)$ by the following push out diagram of abelian groups where $i : \Gamma(A) \to cok(g)$ is the quotient map for the cokernel of g.

$$\begin{array}{ccc} \Gamma(A)\otimes\mathbb{Z}/2\oplus\Gamma(A)\otimes A & \xrightarrow{\bar{q}} & \Gamma_2^2(A) \\ \downarrow{\scriptstyle i\otimes\mathbb{Z}/2\oplus i\otimes A} & & \downarrow{\scriptstyle p} \\ cok(g)\otimes\mathbb{Z}/2\oplus cok(g)\otimes A & \xrightarrow{q} & \Gamma_2^2(g) \end{array}$$

Here the quotient map $\bar{q}$ is defined in (1.1.9). This yields the functor $\Gamma_2^2 : \mathbf{H}(2,4) \to \mathbf{Ab}$.

1.2.6. Proposition: *Let X be a (2,4)-complex with attaching map $g : B \to \Gamma(A)$. Then one has a natural isomorphism*

$$\Gamma_4(X) \;=\; \Gamma_2^2(g).$$

Proof: We have $\Gamma_4 X = image(i_* : \pi_4 M(A,2) \to \pi_4 X)$ and there is a commutative diagram

$$\begin{array}{ccc} \Gamma_2^2(A) & \xrightarrow{\cong} & \pi_4 M(A,2) \\ \downarrow & & \downarrow{\scriptstyle i_*} \\ \Gamma_2^2(g) & \xrightarrow{j} & \pi_4 X \end{array}$$

in which the induced map j is injective. Here we use (1.1.9). Compare also 11.3.4 of [BHH]. □

1.2.7. COROLLARY: *Let X be a (2,4)-Poincaré-complex with attaching map $g : \mathbb{Z} \to \Gamma(A)$ and $A \neq 0$. Then one has natural isomorphisms*

$$\pi_4(X) \cong \Gamma_4(X) \cong \Gamma_2^2(g).$$

Cochran and Habegger [CH] compute $\pi_4(X)$ in (1.2.7) as an abelian group by

$$\Gamma_2^2(g) \cong (\mathbb{Z}/2)^{(r-1)(r+2)/2} \oplus \mathbb{Z}^{r(r-2)(r+2)/3} \tag{1.2.8}$$

where $r = rank(A) > 0$.

1.3. The homotopy category of (2,4)-complexes

The homology functor

$$H_* : \mathbf{CW}(2,4) \to \mathbf{H}(2,4) \tag{1.3.1}$$

carries a (2,4)-complex X to the attaching map $g : B \to \Gamma(A)$ of X. By (1.2.3) we may assume that H_* is actually the identity on objects; more precisely we choose for each object $g : B \to \Gamma(A)$ in $\mathbf{H}(2,4)$ precisely one object C_g in $\mathbf{CW}(2,4)$ and an isomorphism $H_*C_g = g$. In (1.3.1) the category $\mathbf{CW}(2,4)$ is the homotopy category of such (2,4)-complexes C_g and H_* carries C_g to g.

If $f : B' \to \Gamma(A')$ and $g : B \to \Gamma(A)$ are objects in $\mathbf{H}(2,4)$ then the functor H_* yields on morphism sets the surjection

$$H_* : [C_f, C_g] \twoheadrightarrow Mor(f,g) \tag{1.3.2}$$

Where $Mor(f,g) \subset Hom(A',A) \times Hom(B,B')$ is the set of morphisms $f \to g$ in $\mathbf{H}(2,4)$. For $(\xi,\eta) \in Mor(f,g)$ let

$$[C_f, C_g]_{\xi,\eta} \subset [C_f, C_g]$$

be the subset of all elements $F \in [C_f, C_g]$ with $H_*(F) = (\xi, \eta)$. This subset is non empty since (1.3.2) is surjective. Hence $[C_f, C_g]_{\xi,\eta}$ is the fiber of the map H_* in (1.3.2) over the element $(\xi, \eta) \in Mor(f, g)$. We now describe an abelian group $D(\xi, \eta)$ which acts on this fiber.

1.3.3. DEFINITION: For $(\xi, \eta) \in Mor(f, g)$ as above let

$$D(\xi, \eta) = cok(d : Hom(A', cok(g)) \to Hom(B', \Gamma_2^2 g))$$

where d carries $\alpha : A' \to cok(g)$ to the homomorphism

$$d(\alpha) = q(\alpha \otimes \mathbb{Z}/2)\sigma f + q(\alpha \otimes \eta)\tau f.$$

Here q is the map in (1.2.5) and σ and τ are defined in (1.1.2). Hence $d(\alpha)$ is the sum of the composites

$$B' \xrightarrow{f} \Gamma(A') \xrightarrow{\sigma} A' \otimes \mathbb{Z}/2 \xrightarrow{\alpha \otimes \mathbb{Z}/2} cok(g) \otimes \mathbb{Z}/2 \xrightarrow{q} \Gamma_2^2(g),$$

$$B' \xrightarrow{f} \Gamma(A') \xrightarrow{\tau} A' \otimes A' \xrightarrow{\alpha \otimes \eta} cok(g) \otimes A \xrightarrow{q} \Gamma_2^2(g).$$

1.3.4. PROPOSITION: *The abelian group $D(\xi, \eta)$ acts transitively and effectively on the fiber $[C_f, C_g]_{\xi,\eta}$ of the homology H_* in (1.3.2). Compare the proof of (1.3.8) below.*

The action in (1.3.4) has further properties which lead to the notion of natural systems and linear extensions as follows.

1.3.5. DEFINITION: Let $\mathbf{C}$ be a category. A *natural system* of abelian groups on $\mathbf{C}$ is a functor $D : F\mathbf{C} \to \mathbf{Ab}$. Here $F\mathbf{C}$ is the category of factorizations in $\mathbf{C}$, objects in $F\mathbf{C}$ are morphisms $f : A \to B$ in $\mathbf{C}$ and morphisms $(a, b) : f \to g$ are commutative diagrams

$$\begin{array}{ccc} A & \xleftarrow{a} & A' \\ \downarrow f & & \downarrow g \\ B & \xrightarrow{b} & B' \end{array}$$

where bfa is a factorization of g. Composition is defined by $(a', b')(a, b) = (aa', b'b)$. We clearly have $(a, b) = (a, 1)(1, b) = (1, b)(a, 1)$ and we write

$D(f) = D_f$ and $a_* = D(a,1)$, $b^* = D(1,b)$ for the induced maps of the functor D. We say that D is a *bimodule* if $D(f) = D(A,B)$ is given by a bifunctor $D : \mathbf{C}^{op} \times \mathbf{C} \to \mathbf{Ab}$.

1.3.6. EXAMPLE: The groups $D(\xi,\eta)$ in (1.3.3) define a natural system on $\mathbf{H}(2,4)$. For morphisms

$$e \xrightarrow{(\xi',\eta')} f \xrightarrow{(\xi,\eta)} g$$

with $e : B'' \to \Gamma(A'')$ the induced maps

$$\begin{aligned}(\xi,\eta)_* : D(\xi',\eta') &\to D(\xi\xi',\eta\eta'),\\ (\xi',\eta')^* : D(\xi,\eta) &\to D(\xi\xi',\eta\eta')\end{aligned}$$

are induced by $(\xi,\eta)_* = \Gamma_2^2(\eta)_* = Hom(B'',\Gamma_2^2(\xi,\eta))$, $(\xi',\eta')^* = (\xi')^* = Hom(\xi',\Gamma_2^2(g))$. This is an example of a natural system which is not a bimodule.

1.3.7. DEFINITION: Let D be a natural system on the category $\mathbf{C}$. A *linear extension* of the category $\mathbf{C}$ by D, denoted by

$$D \xrightarrow{+} \mathbf{E} \xrightarrow{p} \mathbf{C}\,,$$

is a functor p with the following properties.

(a) $\mathbf{E}$ and $\mathbf{C}$ have the same objects and p is a full functor which is the identity on objects.
(b) For each $f : B \to A$ in $\mathbf{C}$ the abelian group D_f acts transitively and effectively on the subset $p^{-1}(f)$ of morphisms in $\mathbf{E}$. We write $f_0 + \alpha$ for the action of $\alpha \in D_f$ on $f_0 \in p^{-1}(f)$.
(c) The action satisfies the linear distributivity law

$$(f_0+\alpha)(g_0+\beta) = f_0g_0 + f_*\beta + g^*\alpha.$$

Two linear extensions $\mathbf{E}$ and $\mathbf{E}$' are *equivalent* if there is an isomorphism of categories $\epsilon : \mathbf{E} \cong \mathbf{E}'$ with $p\epsilon = p$ and $\epsilon(f_0+\alpha) = \epsilon(f_0)+\alpha$.

The homology functor H_* on (2,4)-complexes in (1.3.1) yields a linear extension of categories as follows.

1.3.8. THEOREM: *Using the homology functor H_* and the natural system D in (1.3.6) and the action in (1.3.4) one obtains a linear extension of categories*

$$D \xrightarrow{+} \mathbf{CW}(2,4) \xrightarrow{H_*} \mathbf{H}(2,4).$$

Proof: Each morphism $(\xi, \eta) : f \to g$ as in (1.2.2)(1) corresponds to a homotopy commutative diagram

$$(1) \qquad \begin{array}{ccc} M(B',3) & \xrightarrow{\xi} & M(B,3) \\ {\scriptstyle f}\downarrow & & \downarrow{\scriptstyle g} \\ M(A',2) & \xrightarrow[\eta]{} & M(A,2) \end{array}$$

in $\mathbf{Top}^*$ where H is a homotopy $H : \eta f \simeq g\xi$. The homotopy H determines a principal map $C(\xi, \eta, H) : C_f \to C_g$ between mapping cones which induces (ξ, η) in homology, compare V.2 in [BAH]. This shows that the functor H_* is full. Let

$$(2) \qquad i_g : M(A,2) \to C_g$$

$$(3) \qquad \nu : C_f \to C_f \vee M(B',4)$$

be the inclusion and the coaction respectively. We define an *action* of

$$(4) \qquad \alpha \in Hom(B', \Gamma_2^2 A) \;=\; [M(B',4), M(A,2)]$$

on $F \in [C_f, C_g]$ by $F + \alpha = (F, i_f\alpha)\nu$. Let $U_F = \{\alpha : F + \alpha = F\}$ be the isotropy group of this action. Then $U_F = I(f, \eta, g)$ depends only on $(\xi, \eta) = H_*(F)$. Now we define the natural system D on $\mathbf{H}(2,4)$ by the quotient group

$$(5) \qquad D(\xi, \eta) \;=\; Hom(B', \Gamma_2^2 A)/I(f, \eta, g).$$

Induced maps for D are defined in the obvious way by naturality in B' and A respectively. We can consider (1.3.8) as a special case of V.7.17 in [BAH]; this also implies that $D(\xi, \eta)$ can be computed by the algebraic description in (1.3.3), compare also 5.12 in [BCU]. □

1.3.9. REMARK: It is well known that extensions of a group G by a G-module D are classified by the second cohomology group $H^2(G,D)$. A "linear extension" $\mathbf{E}$ is the canonical categorical generalization of such extensions of groups and indeed one has the classification

$$\psi : M(\mathbf{C}, D) \cong H^2(\mathbf{C}, D).$$

Here $M(\mathbf{C}, D)$ is the set of equivalence classes of linear extensions of $\mathbf{C}$ by D where we assume $\mathbf{C}$ to be a small category, compare IV.6 in [BAH]. The group $H^n(\mathbf{C}, D)$ is the *cohomology* of $\mathbf{C}$ with coefficients in D which is defined by the nerve of $\mathbf{C}$. We obtain a *representing cocycle* δ_t of the cohomology class $\{\mathbf{E}\} = \psi\mathbf{E}$ as follows. Let t be a splitting function for $p : \mathbf{E} \to \mathbf{C}$ which associates with each morphism f in $\mathbf{C}$ a morphism $f_0 = t(f)$ in $\mathbf{E}$ with $pf_0 = f$. Then t yields the cocycle δ_t by the formula $t(gf) = t(g)t(f) + \delta_t(g,f)$ with $\delta_t(g,f) \in D(gf)$. The cohomology class $\{\mathbf{E}\} = \{\delta_t\}$ is trivial if and only if there is a functor $t : \mathbf{C} \to \mathbf{E}$ with $pt = 1$, in this case $\mathbf{E}$ is a *split* extension of $\mathbf{C}$.

By (1.3.3) and (1.3.6) we have a purely algebraic description of the natural system D on $\mathbf{H}(2,4)$ so that by (1.2.4) the cohomology class

$$\{\mathbf{CW}(2,4)\} \in H^2(\mathbf{H}(2,4), D) \tag{1.3.10}$$

is defined.

1.3.11. REMARK: In V.3.13 in [BAH] we describe three problems for the linear extension **PRIN**. In fact $\mathbf{CW}(2,4)$ is an example of such a category **PRIN** for which we will solve these problems completely. In particular the computation of D in (1.3.3) solves the "isotropy problem" and the computation of the representing cocycle for (1.2.7) solves the "extension problem". We can compute such a cocycle by the algebraic model of the category $\mathbf{CW}(2,4)$ below.

Next we describe groups of automorphisms in a linear extension.

1.3.12. DEFINITION: Let D be a natural system on the category $\mathbf{C}$. For an object A in $\mathbf{C}$ let $Aut_{\mathbf{C}}(A)$ be the group of automorphisms of A in $\mathbf{C}$. Then the abelian group $D_A = D(1_A)$ is a right $Aut_{\mathbf{C}}(A)$-module by defining

$$x^\alpha = (\alpha^{-1})_* \alpha^*(x)$$

for $x \in D_A$ and $\alpha \in Aut(_{\mathbf{C}}(A)$. Given a linear extension $D \xrightarrow{+} \mathbf{E} \xrightarrow{p} \mathbf{C}$

one obtains the group extension

$$0 \longrightarrow D_A \xrightarrow{\ i\ } Aut_{\mathbf{E}}(A) \xrightarrow{\ p\ } Aut_{\mathbf{C}}(A) \longrightarrow 0$$

with $i(x) = 1_A + x$; that is, $Aut_{\mathbf{E}}(A)$ ia an extension of $Aut_{\mathbf{C}}(A)$ by the $Aut_{\mathbf{C}}(A)$-module D_A above. It is well known that such an extension represents an element in

$$\{Aut_{\mathbf{E}}(A)\} \in H^2(Aut_{\mathbf{C}}(A), D_A).$$

This element is the image of the class $\{\mathbf{E}\}$ under the restriction homomorphism

$$H^2(\mathbf{C}, D) \to H^2(Aut_{\mathbf{C}}(A), D_A).$$

We can apply (1.3.12) for the linear extension in (1.3.8). This yields the following result. For a pointed space X let $Aut(X)$ be the subgroup of invertible elements in the monoid $[X, X]$. Then $Aut(X)$ is termed the *group of homotopy equivalences* of X.

1.3.13. PROPOSITION: *Let X be a (2,4)-complex with attaching map $g : B \to \Gamma(A)$. Let $Aut(g)$ be the group of automorphisms of g in $\mathbf{H}(2,4)$ so that by (1.3.12) the $Aut(g)$-module $D(1,1)_g$ is defined where $(1,1)_g$ is the identity of g in $\mathbf{H}(2,4)$. Then the group $Aut(X)$ of homotopy-equivalences of X is part of the extension*

$$0 \to D(1,1)_g \to Aut(X) \twoheadrightarrow Aut(g) \to 0$$

The cohomology class of this extension is the restriction of the class $\{\mathbf{CW}(2,4)\}$.

CHAPTER 2

THE HOMOTOPY CATEGORY OF SIMPLY CONNECTED 4-MANIFOLDS

A (2,4)-Poincaré complex X is a CW-complex of the form

$$X = S^2 \vee \ldots \vee S^2 \cup_f e^4$$

where the attaching map f is given by a unimodular quadratic form on the free abelian group $A = H_2X$; see (1.1.2)(2). By a well known result of Freedman [F] each such (2,4)-Poincaré complex is homotopy equivalent to a simply connected closed topological 4-manifold and vice versa. Therefore the homotopy category $\mathbf{P}(2,4)$ of (2,4)-Poincaré complexes is equivalent to the homotopy category of simply connected closed topological 4-manifolds. The category $\mathbf{P}(2,4)$ is again part of a linear extension of categories. We compute the subcategory $\mathbf{P}(2,4)_0$ of maps $F : X \to Y$ in $\mathbf{P}(2,4)$ with non trivial degree

$$0 \neq F_* = \xi : \ \mathbb{Z} = H_4X \longrightarrow H_4Y = \mathbb{Z}.$$

Compare (2.7.3). This, in particular, yields a result of Cochran and Habegger [CH] on the group $Aut(X)$ of homotopy equivalences of X; see (2.5.2).

2.1. The homotopy category of $(2,4)$-Poincaré complexes

We introduce the *Stiefel-Whitney* class

$$(2.1.1) \qquad w_2 = w_2(X) \in H^2(X, \mathbb{Z}/2) = Hom(H_2X, \mathbb{Z}/2)$$

of a (2,4)-Poincaré complex X. The class w_2 is the unique class which satifies the cup product formula

$$x \cup x = x \cup w_2 \text{ for all } x \in H^2(X, \mathbb{Z}/2) \tag{1}$$

compare (1.2.1)(2). If Z is a basis of $A = H_2X$ and if $(\beta_{m,n} : m, n \in Z)$ is a matrix of the quadratic form $g = b_X[X] \in \Gamma(A)$, see (1.1.2)(2), then the homomorphism $w_2 = w_2^g$ is also given by

$$w_2(m) \equiv \beta_{m,m} \; mod \; 2 \text{ for } m \in Z. \tag{2}$$

We say that the quadratic form $b_X[X]$ is *even* (resp. *odd*) if $w_2 = 0$ (resp. $w_2 \neq 0$). Using (2) we obtain the following equivalences:

$$\begin{aligned} b_X[X] \text{ is even} \quad &\Leftrightarrow \quad w_2 = 0 \\ &\Leftrightarrow \quad \sigma b_X[X] = 0 \\ &\Leftrightarrow \quad \Sigma X \simeq S^5 \vee M(H_2X, 3) \end{aligned}$$

where we use (1.1.7). Let $g : \mathbb{Z} \to \Gamma(A)$ and $f : \mathbb{Z} \to \Gamma(A')$ be objects in **UF** and let (ξ, η) be a morphism in **UF**. Then one has the commutative diagram

$$\begin{array}{ccc} A' & \xrightarrow{\eta} & A \\ {\scriptstyle w_2^f}\downarrow & & \downarrow{\scriptstyle w_2^g} \\ \mathbb{Z}/2 & \xrightarrow{\xi} & \mathbb{Z}/2 \end{array} \tag{3}$$

where ξ is induced by $\xi : \mathbb{Z} \to \mathbb{Z}$. Let $P(w_2^f, \eta) = P$ be the push out in the diagram

$$\begin{array}{ccccc} A' & \xrightarrow{\eta} & A & & \\ {\scriptstyle w_2^f}\downarrow & & \downarrow{\scriptstyle \bar{w}} & \searrow{\scriptstyle w_2^g} & \\ \mathbb{Z}/2 & \xrightarrow{\bar{\eta}} & P & & \\ & \searrow_{\xi} & & \searrow{\scriptstyle \bar{\xi}} & \\ & & & & \mathbb{Z}/2 \end{array} \tag{2.1.2}$$

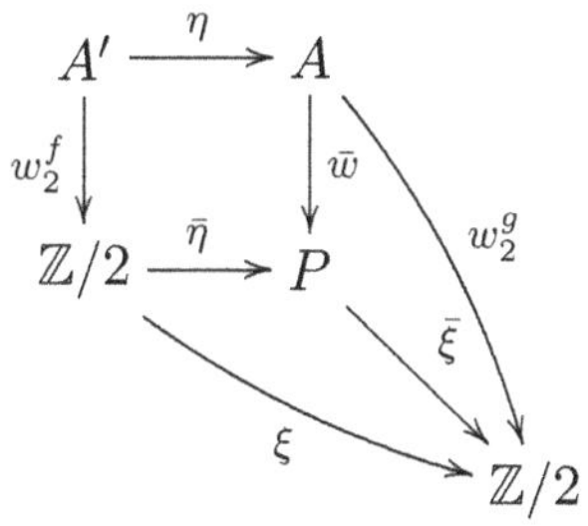

Then $\bar{\eta}, \bar{w}$ yield the quotient map $j = (\bar{\eta}, \bar{w}) : \mathbb{Z}/2 \oplus A \to P = P(w_2^f, \eta)$.

2.1.3. DEFINITION: We define a natural system D on the category **UF** as follows. For $(\xi,\eta) : f \to g$ in **UF** with $f : \mathbb{Z} \to \Gamma(A')$ and $g : \mathbb{Z} \to \Gamma(A)$ and $\xi \in \mathbb{Z}$ let $D(\xi,\eta)$ be given by the push out diagram of abelian groups

$$\begin{array}{ccc} \Gamma(A) \otimes (\mathbb{Z}/2 \oplus A) & \longrightarrow & \Gamma_2^2(A) \\ \downarrow{\scriptstyle i\otimes j} & & \downarrow \\ cok(g) \otimes P(w_2^f, \eta) & \longrightarrow & D(\xi,\eta) \end{array}$$

Here $i : \Gamma(A) \to cok(g)$ and $j : \mathbb{Z}/2 \oplus A \to P(w_2^f,\eta)$ are the quotient maps. For morphisms

$$e \xrightarrow{(\xi',\eta')} f \xrightarrow{(\xi,\eta)} g$$

in **UF** the induced maps

$$\begin{array}{rcl} (\xi,\eta)_* : D(\xi',\eta') & \to & D(\xi\xi',\eta\eta'), \\ (\xi',\eta')^* : D(\xi,\eta) & \to & D(\xi\xi',\eta\eta') \end{array}$$

are induced by $(\xi,\eta)_* = \Gamma_2^2(\eta) : \Gamma_2^2(A') \to \Gamma_2^2(A)$ and $(\xi',\eta')^* = \xi' : \Gamma_2^2(A) \to \Gamma_2^2(A)$. Here ξ' carries $x \in \Gamma_2^2(A)$ to $\xi' \cdot x \in \Gamma_2^2(A)$. One can check that $(\xi,\eta)_*$ and $(\xi',\eta')^*$ are well defined.

2.1.4. PROPOSITION: *The abelian group $D(\xi,\eta)$ in (2.1.3) coincides with $D(\xi,\eta)$ in (1.3.3).*

For the proof of (2.1.4) we show in (2.1.7) that $D(\xi,\eta)$ in (1.3.3) for $(\xi,\eta) : f \to g$ in **UF** is given by

$$D(\xi,\eta) \cong (\Gamma(A) \otimes \mathbb{Z}/2 \oplus \Gamma(A \otimes A))/W$$

where the subgroup W is generated by the elements (i), (ii), (iii) in (1.1.9) and by the elements

(iv) $g(1) \otimes 1,\ g(1) \otimes x,\ x \in A$,
(v) $y \otimes w_2(z) + y \otimes \eta(z),\ y \in \Gamma(A),\ z \in A'$,

where $w_2 = w_2^f$ is defined as in (2.1.1)(2).

We point out that (1.3.2) and (1.3.4) imply the following *classification of maps*. Let X' and X be (2,4)-Poincaré complexes or 1-connected 4-dimensional manifolds with quadratic forms f and g respectively. Moreover let $[X',X]_{\xi,\eta}$ be the subset of $[X', X]$ of all maps inducing (ξ, η) in homology. Then one has a bijection

$$(2.1.5) \qquad D(\xi,\eta) \approx [X', X]_{\xi,\eta}$$

where $D(\xi, \eta)$ is given by (2.1.3). Choosing $F \in [X', X]_{\xi,\eta}$ we obtain a bijection (1.3.4) by mapping $\alpha \in D(\xi, \eta)$ to $F + \alpha$ where we use the action defined in (1.3.8).

For the proof of (2.1.4) we use the following lemma, compare (1.2.1)(5).

2.1.6. LEMMA: *Let $f(1) \in \Gamma(A')$ be unimodular and let $z \in A' = \mathbb{Z} \otimes A'$. Then there is exactly one $x \in Hom(A', \mathbb{Z})$ with $(x \otimes A')\tau f(1) = z$.*

2.1.7. Proof (of (2.1.4)): The generators (i)...(iv) yield $\Gamma_2^2(g)$ in (1.2.5). Therefore it remains to check that the elements (v) generate the image of d in (1.3.3). For this we observe that the group $Hom(A', cok(g))$ is generated by compositions of the form

$$(1) \qquad \alpha : A' \xrightarrow{x} \mathbb{Z} \xrightarrow{y} \Gamma(A) \xrightarrow{i} cok(g).$$

We thus can describe $\varphi = (\alpha \otimes \eta)\tau f$ in (1.3.3) by the composition

$$(2) \qquad \varphi : \mathbb{Z} \xrightarrow{\tau f} A' \otimes A' \xrightarrow{x \otimes A'} \mathbb{Z} \otimes A' \xrightarrow{y \otimes A'} \Gamma(A) \otimes A' \xrightarrow{i \otimes \eta} cok(g) \otimes A.$$

Moreover we use for the computation of $(\alpha \otimes \mathbb{Z}/2)\sigma f$ in (1.3.3) the following equations, where $q_2 : \mathbb{Z} \to \mathbb{Z}/2$ is reduction *mod* 2.

$$(3) \qquad \begin{aligned} (x \otimes \mathbb{Z}/2)\sigma f &= \sigma\Gamma(x)f \\ &= \sigma\tau^{-1}(x \otimes x)\tau f \\ &= q_2(x \otimes x)\tau f \\ &= q_2(x \cup x) \text{ (see (1.2.1)(2))} \\ &= q_2(x \cup w_2) \text{ (see (2.1.1)(1))} \\ &= (q_2 \otimes w_2)(x \otimes A')\tau f. \end{aligned}$$

This shows that $\psi = (\alpha \otimes \mathbb{Z}/2)\sigma f$ is the composition

$$(4) \qquad \psi : \mathbb{Z} \xrightarrow{\tau f} A' \otimes A' \xrightarrow{x \otimes A'} \mathbb{Z} \otimes A' \xrightarrow{y \otimes A'} \Gamma(A) \otimes A \xrightarrow{i \otimes w_2^f} cok(g) \otimes \mathbb{Z}/2.$$

Now let x be determined by $z \in \mathbb{Z} \otimes A'$ as in (2.1.6). Then we get for $d(\alpha) = q(\varphi + \psi)$ in (1.3.3) the equation

$$(5) \qquad d(\alpha)(1) = q(\varphi(1) + \psi(1)) = q(i \otimes \eta + i \otimes w_2)(y \otimes A')(z).$$

Since the group $\Gamma(A) \otimes A'$ is generated by the elemnts $(y \otimes A')(z)$ for $y \in Hom(\mathbb{Z}, \Gamma(A))$ and $z \in \mathbb{Z} \otimes A'$ we thus see that (5) yields the relation (v) in (2.1.4) above. □

We derive from (2.1.4) and (1.3.8) the following result.

2.1.8. THEOREM: *Using the homology functor H_* and the natural system D in (2.1.3) and the action (1.3.4) one obtains a linear extension of categories*

$$D \xrightarrow{+} \mathbf{P}(2,4) \xrightarrow{H_*} \mathbf{UF}.$$

By (1.3.9) we see that the linear extension $\mathbf{P}(2,4)$ represents an element

$$\{\mathbf{P}(2,4)\} \in H^2(\mathbf{UF}, D).$$

The cohomology $\{\mathbf{P}(2,4)\}$ is trivial if and only if there is a functor $s : \mathbf{UF} \to \mathbf{P}(2,4)$ which is a splitting of the homology functor H_*, i.e. $H_* s = id$.

2.2. Maps of non-trivial degree

Let $\mathbf{P}(2,4)_0$ be the following category: objects are (2,4)-Poincaré complexes and morphisms are homotopy classes of maps $F : X \to Y$ with non trivial degree (that is $H_4 F \neq 0$). Similarly let $\mathbf{UF}_0$ be the category of unimodular forms and of maps $(\xi, \eta) : f \to g$ in $\mathbf{UF}$ with $\xi \neq 0$. Then the restriction of the linear extension (2.1.8) yields a linear extension of categories

$$(2.2.1) \qquad D \xrightarrow{+} \mathbf{P}(2,4)_0 \xrightarrow{H_*} \mathbf{UF}_0.$$

Here $D(\xi, \eta)$ is defined for (ξ, η) in $\mathbf{UF}_0$ in the same way as in (2.1.3).

2.2.2. THEOREM: *For $(\xi, \eta) : f \to g$ in $\mathbf{UF}_0$ the abelian group $D(\xi, \eta)_0$ is finite.*

For the proof of (2.2.2) we use the next lemma.

2.2.3. LEMMA: *Let* $f : \mathbb{Z} \to \Gamma(A')$ *and* $g : \mathbb{Z} \to \Gamma(A)$ *and let* $(\xi, \eta) : f \to g$ *be a morphism in* **UF**. *Then the Poincaré duality isomorphisms* ψ_f, ψ_g *are defined as in (1.2.1)(5) and the following diagram commutes:*

$$\begin{array}{ccccc} Hom(A,\mathbb{Z}) & \xleftarrow[\psi_g]{\cong} & A & & \\ \downarrow{\eta^*} & & & \searrow{\xi} & \\ Hom(A',\mathbb{Z}) & \xrightarrow[(\psi_f)^{-1}]{\cong} & A' & \xrightarrow{\eta} & A \end{array}$$

Here $\xi = \xi \otimes A$ *is multiplication by* $\xi \in \mathbb{Z}$.

Proof: Let $\psi_g z = x$ so that $(x \otimes A)\tau g(1) = z$. Then we get

$$\begin{aligned} \xi \cdot z &= (x \otimes A)\tau(\xi \cdot g(1)) \\ &= (x \otimes A)\tau(\Gamma\eta)f(1) \\ &= (x\eta \otimes \eta)\tau f(1) \\ &= \eta(x\eta \otimes A')\tau f(1) \\ &= \eta\psi_f^{-1}(\eta^* x) \\ &= \eta\psi_f^{-1}\eta^*\psi_g(z). \end{aligned}$$

□

The lemma implies that η is surjective if $\xi = 1$ or $\xi = -1$. Moreover η^* is injective if $\xi \neq 0$. For Poincaré complexes X^f and X^g in $\mathbf{P}(2,4)$ corresponding to the forms f and g diagram (2.2.3) corresponds to the following commutative diagram of homology and cohomology groups:

$$\begin{array}{ccc} H^2(X^g) & \xrightarrow{\cap[X^g]} & H_2(X^g) \\ \downarrow{F^*} & & \uparrow{F_*} \\ H^2(X^f) & \xrightarrow{\cap F_*[X^g]} & H_2(X^f) \end{array}$$

Here $F : X^f \to X^g$ induces $(\xi, \eta) : f \to g$ so that $F^* = \eta^*$, $F_* = \eta$, and $F_*[X^g] = \xi[X^f]$.

Proof (of (2.2.2)): We have $D(\xi,\eta) \equiv [X', X]_{\xi,\eta}$ where $D(\xi,\eta)$ admits a quotient map

$$\Gamma(A) \otimes P(w_2^f, \eta) \twoheadrightarrow D(\xi, \eta)$$

by (2.1.3). Here one has an exact sequence

$$\mathbb{Z}/2 \to P(w_2^f, \eta) \to cok(\eta) \to 0$$

where $cok(\eta)$ admits a quotient map $A \otimes (\mathbb{Z}/\xi\mathbb{Z}) \twoheadrightarrow cok(\eta)$ by (2.2.3). □

2.2.4. COROLLARY: *Let X', X be (2,4)-Poincaré complexes. Then the set $[X', X]_{\xi,\eta}$ is finite if $\xi \neq 0$. See (1.3.2)*

Each finite abelian group D has a natural splitting

$$D = \bigoplus_{p \text{ prime}} D^p = \times_{p \text{ prime}} D^p$$

where D^p is the p-primary part of D. This implies that the natural system D on $\mathbf{UF}_0$ also splits as an infinite product.

2.2.5. COROLLARY: *We have natural systems $D^p(\xi, \eta)$ on $\mathbf{UF}_0$ such that*

$$D(\xi, \eta) = \times_{p\ prime} D^p(\xi, \eta)$$

is an isomorphism of natural systems.

Hence the linear extension (2.2.1) represents a cohomology class

$$< \mathbf{P}(2,4) > \in H^2(\mathbf{UF}_0, D) = \times_{p \text{ prime}} H^2(\mathbf{UF}_0, D^p) \tag{2.2.6}$$

so that $< \mathbf{P}(2,4) >$ is completely determined by the coordinates $< \mathbf{P}(2,4) >^p$ in $H^2(\mathbf{UF}_0, D^p)$. We now describe a further splitting of the natural system D on $\mathbf{UF}_0$. For this we need the following natural system.

2.2.7. DEFINITION: Let $(\xi, \eta) : f \to g$ be a morphism in $\mathbf{UF}$ with $f : \mathbb{Z} \to \Gamma(A')$, $g : \mathbb{Z} \to \Gamma(A)$. Then we define $L(\xi, \eta)$ by the push out diagram of abelian groups

$$\begin{array}{ccc}
\Gamma(A)\otimes A & \xrightarrow{W} & L(A,1)_3 \\
\Big\downarrow{\scriptstyle i\otimes p} & & \Big\downarrow \\
cok(g)\otimes cok(\eta) & \xrightarrow{\bar{W}} & L(\xi,\eta)
\end{array}$$

Here $i : \Gamma(A) \to cok(g)$ and $p : A \to cok(\eta)$ are the quotient maps for cokernels and W is defined by $W(\gamma(x) \otimes y) = -[[y,x],x]$ for $x, y \in A$. Induced maps for the natural system L are defined in the same way as in (2.1.3).

2.2.8. LEMMA: *There is a natural exact sequence*

$$\Gamma(A) \otimes \mathbb{Z}/2 \xrightarrow{\bar{i}} D(\xi,\eta) \xrightarrow{\bar{\bar{p}}} L(\xi,\eta) \longrightarrow 0$$

where $\bar{\bar{p}}$ *is split if* $\bar{p} : P(w_2^f,\eta) \twoheadrightarrow cok(\eta)$ *in (2.1.2) is split.*

Proof: There is a push out diagram

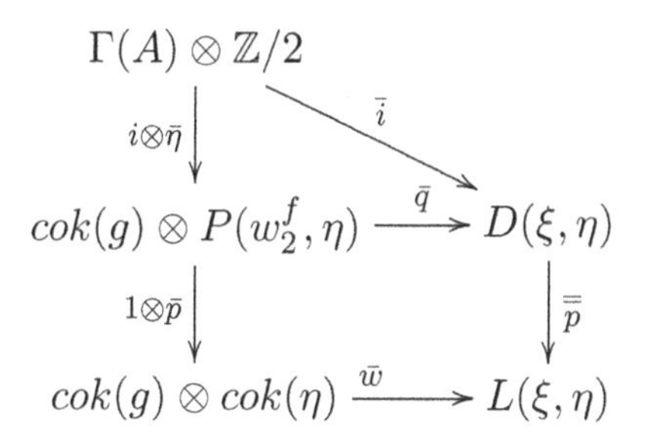

where $\bar{q}$ is given by the push out diagram in (2.1.3). The left hand column is exact since $cok(g)$ is free abelian. Since the push out is also a pull back we obtain exactness of the sequence in (2.2.8). A splitting of $\bar{p}$ induces a splitting of $\bar{\bar{p}}$. □

2.2.9. DEFINITION: Let $(\xi,\eta) : f \to g$ be a morphism in $\mathbf{UF}_0$ with $f : \mathbb{Z} \to \Gamma(A')$ and $g : \mathbb{Z} \to \Gamma(A)$. Then we define a natural system on $\mathbf{UF}_0$ by

$$K(\xi,\eta) = \begin{cases} kernel(w_2^g \otimes \mathbb{Z}/2 : A \otimes \mathbb{Z}/2 \to \mathbb{Z}/2) & \text{if } \xi \text{ is odd,} \\ cokernel(\sigma g : \mathbb{Z} \to \Gamma(A) \to A \otimes \mathbb{Z}/2) & \text{if } w_2^f = 0, \\ 0 & \text{otherwise.} \end{cases}$$

We point out that ξ odd and $w_2^f = 0$ imply $\sigma g = 0$ and $w_2^g = 0$ by (2.1.1) and (2.2.3). Induced maps $(\xi,\eta)^*$ are given via multiplication by ξ and induced maps $(\xi,\eta)_*$ are induced by $\eta \otimes \mathbb{Z}/2$.

2.2.10. THEOREM: *For (ξ, η) in $\mathbf{UF}_0$ one has an isomorphism of natural systems*

$$D(\xi, \eta) \cong L(\xi, \eta) \oplus K(\xi, \eta).$$

For the proof of theorem (2.2.10) we need the following lemmas.

2.2.11. LEMMA: *There is a push out diagram*

$$\begin{array}{ccc} \Gamma(A) \otimes (\mathbb{Z}/2 \oplus A) & \xrightarrow{q} & \Gamma(A) \otimes \mathbb{Z}/2 \oplus L(A,1)_3 \\ {\scriptstyle 1\otimes(1,0)} \downarrow & & \downarrow {\scriptstyle (\sigma,0)} \\ \Gamma(A) \otimes \mathbb{Z}/2 & \xrightarrow{\sigma} & A \otimes \mathbb{Z}/2 \end{array}$$

Proof: We have to show that the kernel of σ is $(1 \otimes (1,0))(R(A))$ where $R(A)$ denotes the relations in (1.1.9). But this is obvious, since the kernel of σ consists of elements $[x, y] \otimes 1$. □

2.2.12. LEMMA: *Let A be a free abelian group and let $w : A \to \mathbb{Z}/2k$ be a surjective homomorphism, $k \geq 1$. We define*

$$K(w) = \begin{cases} 0 & \text{if } k \text{ is even,} \\ kernel(w \otimes \mathbb{Z}/2 : A \otimes \mathbb{Z}/2 \to \mathbb{Z}/2) & \text{if } k \text{ is odd.} \end{cases}$$

and we associate with w the diagram below where $i : \mathbb{Z}/2 \to \mathbb{Z}/2k$ with $i(1) = k$ is the inclusion and where q is the quotient map in (1.1.9). This diagram is a push out diagram.

$$\begin{array}{ccc} \Gamma(A) \otimes (\mathbb{Z}/2 \oplus A) & \xrightarrow{q} & \Gamma(A) \otimes \mathbb{Z}/2 \oplus L(A,1)_3 \\ {\scriptstyle 1\otimes(i,w)} \downarrow & & \downarrow {\scriptstyle v} \\ \Gamma(A) \otimes \mathbb{Z}/2k & \xrightarrow{u} & K(w) \end{array}$$

If k is odd we define u and v as follows with $x, y, z \in A$.

$$(1) \qquad v([[x,y],z]) = (x \otimes w(y) + y \otimes w(x)) \cdot w(z),$$

$$(2) \qquad v(\gamma(a) \otimes 1) = u(\gamma(x) \otimes 1) = x \otimes 1 + x \otimes w(x).$$

Here w is the composite $A \to \mathbb{Z}/2k \twoheadrightarrow \mathbb{Z}/2$ given by w above. The formulas imply

$$(3) \qquad u([x,y]\otimes 1) \;=\; x\otimes w(y)+y\otimes w(x).$$

By the universal property of γ in (1.1.1) we see that u is well defined. Moreover the generatorss (1.1.8)(a),(b),(c) show that v is well defined on $L(A,1)_3$. In addition it is easy to see that v and u both map to the kernel of $w\otimes\mathbb{Z}/2$. Next we check that the diagram above commutes. This is clear for elements in $\Gamma(A)\otimes\mathbb{Z}/2$ by (2). Moreover for $\gamma(x)\otimes y\in\Gamma(A)\otimes A$ we get

$$\begin{aligned} vq(\gamma(x)\otimes y) &= v([x,y]\otimes 1-[[y,x],x]) \\ &= x\otimes w(y)+y\otimes w(x)+(y\otimes w(x)+x\otimes w(y))\cdot w(x), \end{aligned}$$

$$\begin{aligned} u(1\otimes w_2^g)(\gamma(x)\otimes y) &= u(\gamma(x)\otimes w(y)) \\ &= (x\otimes 1+x\otimes w(x))\cdot w(y) \\ &= x\otimes w(y)+x\otimes w(x)\cdot w(y). \end{aligned}$$

Proof (of (2.2.12)): The kernel of q in the diagram is given by the generators of $R(A)$ in (1.1.9). We consider the composite

$$\lambda: R(A)(\subset\Gamma(A)\otimes(\mathbb{Z}/2\oplus A)) \xrightarrow{1\otimes(i,w)} \Gamma(A)\otimes\mathbb{Z}/2k$$

so that the push out of the diagram coincides with *cokernel*(λ). We write $x\sim 0$ if $x\in\lambda(R(A))$ Then $\lambda(R(A))$ is generated by the following elements:

$$[x,y]\otimes wz+[z,x]\otimes wy+[y,z]\otimes wx,$$

$$\gamma(x)\otimes wx,$$

$$[x,y]\otimes k+\gamma(x)\otimes wy+[y,x]\otimes wx.$$

We can choose a basis $e_1,\dots,e_n$ of A with $w(e_1)=1$ and $w(e_i)=0$ for $i>1$. Then we get the following relations:

$$[e_i, e_j] \otimes we_1 + [e_1, e_i] \otimes we_j + [e_j, e_1] \otimes we_i = [e_i, e_j] \otimes 1 \sim 0 \text{ for } 1 < i < j,$$

$$[e_1, e_j] \otimes we_1 + [e_1, e_1] \otimes we_j + [e_j, e_1] \otimes we_1 = 2[e_1, e_j] \otimes 1 \sim 0 \text{ for } 1 < j,$$

$$\gamma(e_1) \otimes we_1 = \gamma(e_1) \otimes 1 \sim 0,$$

$$\gamma(e_i) \otimes we_1 + [e_i, e_1] \otimes k + [e_1, e_i] \otimes we_i = \gamma(e_i) \otimes 1 + [e_1, e_i] \otimes k \sim 0 \quad \text{for } 1 < i$$

Hence the push out *cokernel*(λ) is a $\mathbb{Z}/2$-vector space generated by $[e_1, e_j] \otimes 1$ with $1 < j$. If k is even then also $\gamma(e_i) \sim 0$ since $[e_1, e_i] \otimes k \sim 0$. Moreover we get in this case for $1 < i$

$$\begin{aligned} & [e_1, e_i] \otimes k + \gamma(e_1) \otimes we_i + [e_i, e_1] \otimes we_1 \\ = \quad & [e_1, e_i] \otimes k + [e_1, e_i] \otimes 1 \sim [e_1, e_i] \otimes 1 \sim 0, \end{aligned}$$

so that the push out is trivial if k is even. If k is odd we see that

$$u([e_1, e_i] \otimes 1) \;=\; e_i \otimes 1$$

for $i > 1$. This shows by the commutative diagram above that the push out $K(w)$ coincides with *kernel*$(w \otimes \mathbb{Z}/2)$. □

2.2.13. LEMMA: *Assume in (2.2.12) we have $k = 1$ and $w = w_2^g : A \to \mathbb{Z}/2$ is given by the Stiefel-Whitney class of the unimodular form $g(1) \in \Gamma(A)$. Then we have*

$$u(g(1) \otimes 1) \;=\; 0.$$

Proof: For the matrix β_{ij} of $g(1)$ we have

$$g(1) \otimes 1 = (\sum_i \gamma(e_i) \otimes \beta_{ii}) + (\sum_{i<j} [e_i, e_j] \otimes \beta_{ij})$$

where $e_1, \ldots, e_n$ are a basis as in the proof of (2.2.12) with $w(e_1) = 1 \equiv \beta_{11}$ *mod* 2 and $w(e_i) = 0 \equiv \beta_{ii}$ *mod* 2 for $i > 1$. Hence by the relations in the proof of (2.2.12) we get

$$g(1) \otimes 1 \sim \gamma(e_1) \otimes 1 + \sum_{1<j} [e_1, e_j] \otimes \beta_{ij}.$$

Here (2.1.1)(1) implies that $e_i^* \cup e_i^* = e_i^* \cup w_2$ where $w_2 = e_1^*$. Hence $\beta_{ii} \equiv \beta_{1i}\ mod\ 2$ where $\beta_{ii} \equiv 0\ mod\ 2$ for $i > 1$. Hence we get

$$g(1) \otimes 1 \sim \gamma(e_1) \otimes 1 \sim 0.$$

This proves the lemma above. □

2.2.14. Proof (of 2.2.10): We consider several cases. If $w_2^f = 0$ there is a commutative diagram

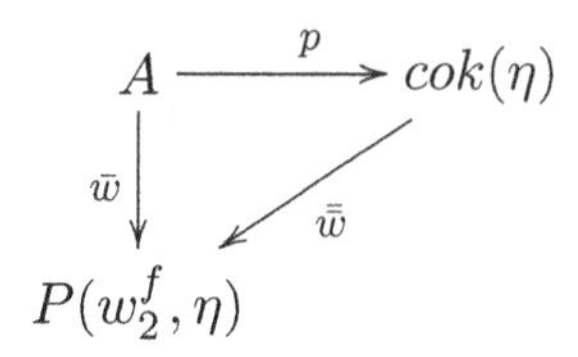

where $\bar{w}$ is a splitting of $\bar{p} : P(w_2^f, \eta) \to cok(\eta)$. This splitting yields a retraction $r : P(w_2^f, \eta) \to \mathbb{Z}/2$ of $\bar{\eta}$ with $r\bar{\bar{w}} = 0$. Hence we obtain the cokernel of the splitting $L(\xi, \eta) \to D(\xi, \eta)$ induced by $\bar{\bar{w}}$ by the following push out diagrams:

$$\begin{array}{ccc}
\Gamma(A) \otimes (\mathbb{Z}/2 \oplus A) & \xrightarrow{q} & \Gamma(A) \otimes \mathbb{Z}/2 \oplus L(A,1)_3 \\
\downarrow{\scriptstyle 1\otimes(1,0)} & & \downarrow \\
\Gamma(A) \otimes \mathbb{Z}/2 & \xrightarrow{\sigma} & A \otimes \mathbb{Z}/2 \\
\downarrow & & \downarrow \\
cok(g) \otimes \mathbb{Z}/2 & \longrightarrow & cok(\sigma g)
\end{array}$$

It is easy to see that the bottom part of the diagram is a push out and by (2.2.11) the top part of the diagram is a push out. This proves theorem(2.2.7) if $w_2^f = 0$.

Next we assume ξ is odd. In this case the map $\bar{\xi}$ in (2.1.2) is a retraction of $\bar{\eta}$ which induces a splitting of $P(w_2^f, \eta) \to cok(\eta)$. The cokernel of the induced splitting $L(\xi, \eta) \to D(\xi, \eta)$ is given by $K(w)$ in the push out diagram (2.2.12) with $w = w_2^g \neq 0$ and $k = 1$. This proves theorem (2.2.10) if ξ is odd and $w_2^g \neq 0$. If ξ is odd and $w_2^g = 0$ we have $\sigma(g(1)) = 0$ and hence in this case the cokernel of $L(\xi, \eta) \to D(\xi, \eta)$ is given by the push out (2.2.11). This proves theorem (2.2.10) if ξ is odd.

Next we consider the case $w_2^f \neq 0$ and ξ even, $\xi \neq 0$. If $\bar{\eta} = 0$ then $P(w_2^f, \eta) = cok(\eta)$ and hence $D(\xi, \eta) = L(\xi, \eta)$. If $\bar{\eta} \neq 0$ and is non-split we can find a commutaive diagram

$$\begin{array}{ccc} \mathbb{Z}/2 & \stackrel{\bar{\eta}}{\rightarrowtail} & P(w_2^f, \eta) \\ \| & & \downarrow t \\ \mathbb{Z}/2 & \stackrel{i}{\rightarrowtail} & \mathbb{Z}/2k \end{array}$$

where t has a splitting s with $si = \bar{\eta}$ and where k is even. Then (2.2.12) with $w = t\bar{w} : A \to P(w_2^f, \eta) \to \mathbb{Z}/2k$ shows that the map $\Gamma(A) \otimes \mathbb{Z}/2 \to D(\xi, \eta)$ in (2.2.8) is trivial. This shows that $D(\xi, \eta) = L(\xi, \eta)$.

Finally we consider $\bar{\eta} \neq 0$ and $\bar{\eta}$ split. This case, however, implies by (2.2.15) below that ξ is odd contradicting the assumption above that ξ is even. This completes the proof of (2.2.10). □

2.2.15. Lemma: *Let $(\xi, \eta) : f \to g$ be a morphism in $\mathbf{UF}_0$. If $w_2^f \neq 0$ and $\bar{\eta}$ split with $\bar{\eta} : \mathbb{Z}/2 \to P(w_2^f, \eta)$ then ξ is odd.*

Proof: Since $\xi \neq 0$ we see that $rank(A') \geq rank(A)$. The push out diagram

$$\begin{array}{ccc} A' \otimes \mathbb{Z}/2 & \xrightarrow{\eta \otimes \mathbb{Z}/2} & A \otimes \mathbb{Z}/2 \\ {\scriptstyle w_2^f \otimes \mathbb{Z}/2} \downarrow & & \downarrow {\scriptstyle \bar{w} \otimes \mathbb{Z}/2} \\ \mathbb{Z}/2 & \stackrel{\bar{\eta} \otimes \mathbb{Z}/2}{\rightarrowtail} & P(w_2^f, \eta) \otimes \mathbb{Z}/2 \end{array}$$

is also a pull back diagram so that $ker(\eta \otimes \mathbb{Z}/2) = 0$; this implies $rank(A') = rank(A)$. Moreover $\eta \otimes \mathbb{Z}/2$ is an isomorphism and $cok(\eta) \otimes \mathbb{Z}/2 = 0$. We can find a basis $e_1', \ldots, e_n'$ of A' and a basis $e_1, \ldots, e_n$ of A with $\eta e_1' = \eta_i \cdot e_i$ with $\eta_i \in \mathbb{Z}$. For the matrices (β_{ij}') of $f(1)$ and (β_{ij}) of $g(1)$ we get

$$\eta_i \eta_j \beta_{ij}' = \xi \beta_{ij}.$$

Applying the determinant shows

$$(\eta_1 \ldots \eta_n)^2 = \pm \xi^n$$

since $det(\beta_{ij}') = \pm 1 = det(\beta_{ij})$. Since $\eta_1, \ldots, \eta_n$ are odd we see that ξ is odd. □

2.3. The double suspension

The natural system K on $\mathbf{UF}_0$ defined in (2.2.9) admits a further splitting as follows. Let $(\xi,\eta): f \to g$ be a morphism in $\mathbf{UF}_0$ with $f: \mathbb{Z} \to \Gamma(A')$ and $g: \mathbb{Z} \to \Gamma(A)$. Then the *stable part* of K is given by the natural system K^{st} on $\mathbf{UF}_0$ with

$$K^{st}(\xi,\eta) = \begin{cases} coker(\sigma g : \mathbb{Z} \to \Gamma(A) \to A \otimes \mathbb{Z}/2) & \text{if } w_2^f = 0, \\ 0 & \text{otherwise.} \end{cases} \tag{2.3.1}$$

The *unstable part* of K is the natural system K^u on $\mathbf{UF}_0$ given by

$$K^u(\xi,\eta) = \begin{cases} ker(w_2^g \otimes \mathbb{Z}/2 : A \otimes \mathbb{Z}/2 \to \mathbb{Z}/2) & w_2^f \neq 0 \text{ and } \xi \text{ odd}, \\ 0 & \text{otherwise.} \end{cases} \tag{2.3.2}$$

We point out that $w_2^f \neq 0$ and ξ odd imply that $w_2^g \neq 0$. One can check that K^{st} and K^u are well defined natural systems and that there is a canonical isomorphism of natural systems

$$K(\xi,\eta) \;=\; K^u(\xi,\eta) \oplus K^{st}(\xi,\eta). \tag{2.3.3}$$

Hence by (2.2.10) we obtain an isomorphism

$$D(\xi,\eta) \;=\; L(\xi,\eta) \oplus K^u(\xi,\eta) \oplus K^{st}(\xi,\eta) \tag{1}$$

of natural systems on $\mathbf{UF}_0$. Hence the cohomology class $< \mathbf{P}(2,4) >$ in (2.2.6) is an element in the direct sum given by the coordinates $< \mathbf{P}^L(2,4) >$, $< \mathbf{P}^u(2,4) >$ and $< \mathbf{P}^{st}(2,4) >$ respectively:

$$< \mathbf{P}(2,4) > \in H^2(\mathbf{UF}_0, D) = H^2(\mathbf{UF}_0, L) \oplus H^2(\mathbf{UF}_0, K^u) \oplus H^2(\mathbf{UF}_0, K^{st}). \tag{2}$$

In addition we can use the splitting with respect to primary parts in (2.2.5) which shows that one has isomorphisms of natural systems on $\mathbf{UF}_0$:

$$\begin{cases} D^p = L^p & \text{for } p \text{ odd}, \\ D^2 = L^2 \oplus K^u \oplus K^{st}. & \end{cases} \tag{3}$$

Hence $< \mathbf{P}(2,4) >^2$ has accordingly the coordinates $< \mathbf{P}^L(2,4) >^2$, $< \mathbf{P}^u(2,4) >$ and $< \mathbf{P}^{st}(2,4) >$.

We now consider the double suspension functor Σ^2 on $\mathbf{P}(2,4)$. For a morphism $(\xi,\eta) : f \to g$ in $\mathbf{UF}$ we have the mapping cones C_f and C_g as in (1.3.2). The double suspension yields a natural map

$$\Sigma^2 : [C_f, C_g]_{\xi,\eta} \longrightarrow [\Sigma^2 C_f, \Sigma^2 C_g]_{\xi,\eta} \tag{2.3.4}$$

where the right hand side denotes the subset of $[\Sigma^2 C_f, \Sigma^2 C_g]$ consisting of all elements F which induce

$$\xi : \mathbb{Z} = H_6\Sigma^2 C_f \xrightarrow{F_*} H_6\Sigma^2 C_g = \mathbb{Z},$$
$$\eta : A' = H_4\Sigma^2 C_f \xrightarrow{F_*} H_4\Sigma^2 C_g = A.$$

For the next result compare V.7.17 in [BAH].

2.3.5. PROPOSITION: *For $(\xi,\eta) : f \to g$ in $\mathbf{UF}_0$ the group $K^{st}(\xi,\eta)$ acts transitively and effectively on the set $[\Sigma^2 C_f, \Sigma^2 C_g]_{\xi,\eta}$ and one has a commutative diagram where the projection is given by (2.3.3)(1).*

$$\begin{array}{ccc} D(\xi,\eta) & \xrightarrow{pr} & K^{st}(\xi,\eta) \\ {\scriptstyle F_+}\downarrow \equiv & & \equiv \downarrow {\scriptstyle (\Sigma^2 F)_+} \\ [C_f, C_g]_{\xi,\eta} & \xrightarrow{\Sigma^2} & [\Sigma^2 C_f, \Sigma^2 C_g]_{\xi,\eta} \end{array}$$

Given $F \in [C_f, C_g]_{\xi,\eta}$ the bijection F_+ carries $\alpha \in D(\xi,\eta)$ to $F_+(\alpha) = F + \alpha$; see (2.1.5). Similarly the bijection $(\Sigma^2 F)_+$ carries $\beta \in K^u(\xi,\eta)$ to $(\Sigma^2 F)_+(\beta) = (\Sigma^2 F) + \beta$. We point out that (2.3.5) implies

2.3.6. COROLLARY: *For $(\xi,\eta) : f \to g$ in $\mathbf{UF}_0$ the double suspension*

$$\Sigma^2 : [C_f, C_g]_{\xi,\eta} \longrightarrow [\Sigma^2 C_f, \Sigma^2 C_g]_{\xi,\eta}$$

is surjective.

2.3.7. DEFINITION: We define the *stable category* $\mathbf{P}^{st}(2,4)$ of simply connected 4-dimensional Poincaré complexes as follows. Objects are mapping cones C_f with f in $\mathbf{UF}_0$ and morphisms $C_f \to C_g$ are triples (ξ, η, F) with $(\xi, \eta) : f \to g$ in $\mathbf{UF}$ and $F \in [\Sigma^2 C_f, \Sigma^2 C_g]_{\xi,\eta}$. By (2.3.5) one has a commutative diagram of linear extensions

$$\begin{array}{ccc} D & \xrightarrow{pr} & K^{st} \\ \downarrow & & \downarrow \\ \mathbf{P}(2,4)_0 & \xrightarrow{\Sigma^2} & \mathbf{P}^{st}(2,4)_0 \\ \downarrow & & \downarrow \\ \mathbf{UF}_0 & \xrightarrow{id} & \mathbf{UF}_0 \end{array}$$

where id is the identity functor. The class $< \mathbf{P}^{st}(2,4) >$ of the extension $\mathbf{P}^{st}(2,4)$ is thus the coordinate of $< \mathbf{P}(2,4) >$ in (2.3.3)(2).

Unsöld [U] computes the following algebraic description of the category $\mathbf{P}^{st}(2,4)$.

2.3.8. DEFINITION: Let $\mathbf{S}(2,4)_0$ be the following category. Objects are objects $g : \mathbb{Z} \to \Gamma(A)$ in $\mathbf{UF}_0$ for which we choose an exact sequence

$$0 \longrightarrow cok(\sigma g) \xrightarrow{j} \pi \longrightarrow \mathbb{Z} \xrightarrow{\sigma g} A \otimes \mathbb{Z}/2 \longrightarrow cok(\sigma g) \longrightarrow 0.$$

and a homomorphism T such that the diagram

$$\begin{array}{ccccc} A & \xrightarrow{\cdot 2} & ker(ir) & \twoheadrightarrow & im(\sigma g) \\ {\scriptstyle ir}\downarrow & & \downarrow{\scriptstyle T} & & \downarrow{\scriptstyle \Omega} \\ cok(\sigma g) & \xrightarrow{\bar{j}} & \pi \otimes \mathbb{Z}/2 & \twoheadrightarrow & \mathbb{Z}/2 \end{array}$$

commutes. Here ir is the composite of quotient maps $A \to A \otimes \mathbb{Z}/2 \to cok(\sigma g)$ and $\cdot 2$ is given by multiplication by 2. Moreover $\bar{j}$ is induced by j in the sequence and Ω is the unique injective homomorphism. The rows of the diagram are exact. Morphisms $f \to g$ in $\mathbf{S}(2,4)_0$ are triples (ξ, η, λ) where $(\xi, \eta) : f \to g$ in $\mathbf{UF}_0$ and $\lambda : \pi' \to \pi$ are morphisms compatible with the sequence and the diagram above, in the sense that the following diagrams commute:

$$\begin{array}{ccccccc} cok(\sigma f) & \longrightarrow & \pi' & \longrightarrow & \mathbb{Z} & \xrightarrow{\sigma f} & A' \otimes \mathbb{Z}/2 \\ {\scriptstyle \eta \otimes \mathbb{Z}/2}\downarrow & & \downarrow{\scriptstyle \lambda} & & \downarrow{\scriptstyle \xi} & & \downarrow{\scriptstyle \eta \otimes \mathbb{Z}/2} \\ cok(\sigma g) & \longrightarrow & \pi & \longrightarrow & \mathbb{Z} & \xrightarrow{\sigma g} & A \otimes \mathbb{Z}/2 \end{array}$$

$$\begin{array}{ccccc} A' \supset & ker(i'r') & \xrightarrow{T'} & \pi' \otimes \mathbb{Z}/2 \\ \eta \downarrow & \downarrow & & \downarrow \lambda\otimes\mathbb{Z}/2 \\ A \supset & ker(ir) & \xrightarrow{T} & \pi \otimes \mathbb{Z}/2 \end{array}$$

The following result is due to Unsöld [U].

2.3.9. THEOREM: *There is an isomorphism of categories*

$$\mathbf{P}^{st}(2,4)_0 \;=\; \mathbf{S}(2,4)_0$$

compatible with the obvious functors to $\mathbf{UF}_0$.

We have a linear extension

$$K^{st} \xrightarrow{+} \mathbf{S}(2,4)_0 \longrightarrow \mathbf{UF}_0$$

isomorphic to the extension for $\mathbf{P}^{st}(2,4)_0$ in (2.3.7). Here the action $+$ is defined by $(\xi,\eta,\lambda)+\alpha = (\xi,\eta,\lambda+j\alpha p')$ where $j : cok(\sigma g) \to \pi$ and $p' : \pi' \to \mathbb{Z}$ is the surjection induced by $\pi' \to \mathbb{Z}$ in the sequence for f.

Though we have an explicit algebraic category $\mathbf{S}(2,4)_0$ it is not easy to see whether the functor $\mathbf{S}(2,4)_0 \to \mathbf{UF}_0$ admits a splitting functor or not. If there is no such splitting functor then the class $< \mathbf{P}^{st}(2,4) >$ is non trivial.

Recall that a form f is odd (resp. even) if $w_2^f \neq 0$ (resp. $w_2^f = 0$). Accordingly we write $\mathbf{C}^{odd}$ (resp. $\mathbf{C}^{even}$) for the full subcategory of $\mathbf{C}$ consisting of odd (resp. even) objects.

2.3.10. COROLLARY: *One has an equivalence of categories*

$$\mathbf{P}^{st}(2,4)_0^{odd} \;=\; \mathbf{UF}_0^{odd}$$

and one has a split extension of categories

$$K^{st} \to \mathbf{P}^{st}(2,4)_0^{even} \to \mathbf{UF}_0^{even}.$$

Proof: We observe that for even forms g the group $im(\sigma g) = 0$ is trivial so that T in (2.3.8) is determined by $\bar{j}ir$. Hence choosing for each g a section $s_g : \mathbb{Z} \to \pi$ of $\pi \to \mathbb{Z}$ we obtain unique maps $\lambda : \pi' \to \pi$ compatible with the section (i.e. $\lambda s_f = s_g$) and a splitting functor carries (ξ, η) to (ξ, η, λ). □

2.4. The odd part and the signature obstruction

Using the splitting $D = L \oplus K^u \oplus K^{st}$ of the natural system D on $\mathbf{UF}_0$ in (2.3.3)(1) we define a natural equivalence relation $\sim$ on $\mathbf{P}(2,4)_0$ as follows. For maps $F, G : C_f \to C_g$ in $\mathbf{P}(2,4)_0$ we write $F \sim G$ if F and G induce the same morphism in homology, i.e. $H_*(F) = H_*(G) = (\xi, \eta) : f \to g$ and if there is

$$\alpha \in L(\xi, \eta) \oplus K^{st}(\xi, \eta)$$

with $F + \alpha = G$. One readily checks that this is a natural equivalence relation so that the quotient category

$$\mathbf{P}^u(2,4)_0 = \mathbf{P}(2,4)_0/\sim = \mathbf{P}(2,4)_0/(L \oplus K^{st})$$

is defined. One has a linear extension

$$K^u \to \mathbf{P}^u(2,4)_0 \to \mathbf{UF}_0 \tag{2.4.1}$$

which represents the cohomolgy class $< \mathbf{P}^u(2,4)_0 >$ in (2.3.3)(3).

2.4.2. DEFINITION: Let $\mathbf{P}(odd)$ and $\mathbf{UF}(odd)$ be the following subcategories of $\mathbf{P}^u(2,4)_0$ and $\mathbf{UF}_0$ respectively. Objects are forms f with $w_2^f \neq 0$ and morphisms are maps with odd degree ξ.

According to the definition of K^u in (2.3.2) we have $K^u(\xi, \eta) \neq 0$ if and only if (ξ, η) is a morphism in the odd part $\mathbf{UF}(odd)$. Moreover we have an inclusion of linear extensions where K^u_{odd} is the restriction of K^u to $\mathbf{UF}(odd)$.

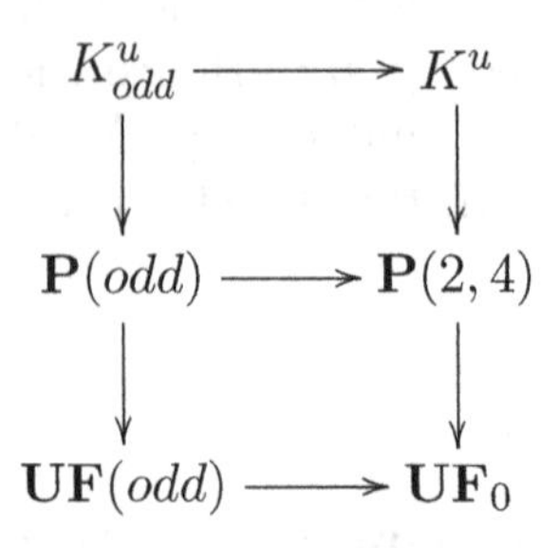

2.4.3. LEMMA: *The linear extension for* $\mathbf{P}(2,4)$ *is determined by* $\mathbf{P}(odd)$.

Proof: For this we observe that for a composite $(\xi',\eta')(\xi,\eta)$ of morphisms in $\mathbf{UF}_0$ the following holds. If (ξ',η') is not in $\mathbf{UF}(odd)$ or if (ξ,η) is not in $\mathbf{UF}(odd)$ then also the composite $(\xi',\eta')(\xi,\eta)$ is not in $\mathbf{UF}(odd)$. □

We now describe an algebraic category equivalent to $\mathbf{P}(odd)$. For this we need the following property of the *signature* τ_f of the form f, see [HNK].

2.4.4. LEMMA: *For each* f *in* $\mathbf{UF}$ *the signature* $\tau_f \in \mathbb{Z}$ *is defined. If* $(\xi,\eta) : f \to g$ *is a morphism in* $\mathbf{UF}(odd)$ *then* $\tau_f \equiv \xi\tau_g$ *mod* 8.

Proof: For $g : \mathbb{Z} \to \Gamma(A')$ we can choose a commutative diagram

(1)
$$\begin{array}{ccc} A\otimes\mathbb{Z}/2 & \twoheadleftarrow & A \\ {\scriptstyle w_2^g}\downarrow & & \downarrow{\scriptstyle v^g} \\ \mathbb{Z}/2 & \twoheadleftarrow & \mathbb{Z} \end{array}$$

where the horizontal arrows are the quotient maps. Then v^g is called a *characteristic element* of the form g since by (2.1.1)(1) we have $x \cup x \equiv x \cup v^g$ *mod* 2 for all $x : A \to \mathbb{Z}$. By 6.3 [HKN] we know

(2)
$$\Gamma(v^g)(g(1)) \;=\; v^g \cup v^g \equiv \tau_g \ mod 8.$$

If $(\xi,\eta) : f \to g$ is a morphism in $\mathbf{UF}(odd)$ then (2.1.1)(3) shows that $v^f = v^g\eta$ is a characteristic element of f so that also $\Gamma(v^f)(f(1)) \equiv \tau_f$ *mod*8. On the other hand we have *mod*8

$$\begin{aligned} \tau_f \equiv \Gamma(v^f)(f(1)) &= \Gamma(v^g\eta)(f(1)) &&= \Gamma(v^g)\Gamma(\eta)(f(1)) \\ &= \Gamma(v^g)\xi g(1) &&\equiv \xi\tau_g \end{aligned}$$

□

Lemma (2.4.4) shows that the following diagram commutes for $(\xi, \eta) : f \to g$ in $\mathbf{UF}(odd)$.

(2.4.5)

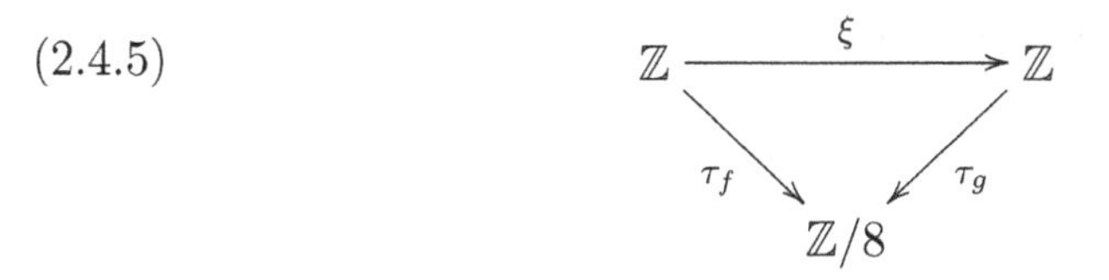

Here τ_f is the homomorphism which carries $1 \in \mathbb{Z}$ to $\{\tau_f\} \in \mathbb{Z}/8$. We now are ready for the definition of the algebraic model for $\mathbf{P}(odd)$.

2.4.6. DEFINITION: Let $\mathbf{S}(odd)$ be the following category. Objects are unimodular forms $g : \mathbb{Z} \to \Gamma(A)$ with $w_2^g \neq 0$. We choose for g a commutative diagram with exact rows

(1)
$$\begin{array}{ccccccccc} 0 & \longrightarrow & A \otimes /Z/2 & \xrightarrow{i} & \Omega(g) & \xrightarrow{p} & \mathbb{Z} & \longrightarrow & 0 \\ & & \downarrow{\scriptstyle w_2^g \otimes \mathbb{Z}/2} & & \downarrow{\scriptstyle \alpha_g} & & \downarrow{\scriptstyle \tau_g} & & \\ 0 & \longrightarrow & \mathbb{Z}/2 & \longrightarrow & \mathbb{Z}/16 & \longrightarrow & \mathbb{Z}/8 & \longrightarrow & 0 \end{array}$$

Morphisms $f \to g$ in $\mathbf{S}(odd)$ are triples (ξ, η, λ) where $(\xi, \eta) : f \to g$ is a morphism in $\mathbf{UF}(odd)$, i.e. ξ is odd, and $\lambda : \Omega(f) \to \Omega(g)$ is a homomorphism such that the diagram

(2)
$$\begin{array}{ccccc} A' \otimes \mathbb{Z}/2 & \longrightarrow & \Omega(f) & \longrightarrow & \mathbb{Z} \\ \downarrow{\scriptstyle \eta \otimes 1} & & \downarrow{\scriptstyle \lambda} & & \downarrow{\scriptstyle \xi} \\ A \otimes \mathbb{Z}/2 & \longrightarrow & \Omega(g) & \longrightarrow & \mathbb{Z} \end{array}$$

commutes with $\alpha_g \lambda = \alpha_f$. That is (ξ, η, λ) is a morphism of the top row of (1) over the bottom row. In fact, we have $w_2^g(\eta \otimes 1) = w_2^f$ and $\tau_g \xi = \tau_f$ by (2.4.5).

We have a linear extension

(2.4.7)
$$K_{odd}^u \longrightarrow \mathbf{S}(odd) \xrightarrow{p} \mathbf{UF}(odd)$$

as follows. The functor p carries (ξ, η, λ) to (ξ, η) and the action of $\alpha \in K^u_{odd}(\xi, \eta) = kernel(w_2^g \otimes \mathbb{Z}/2)$ is defined by

$$(\xi, \eta, \lambda) + \alpha \;=\; (\xi, \eta, \lambda + i\alpha p)$$

where i and p are the maps in the top row of (2.4.6)(1).

2.4.8. THEOREM: *There is an equivalence*

$$\mathbf{P}(odd) \;=\; \mathbf{S}(odd)$$

of linear extensions (2.4.3) and (2.4.7).

Proof: We use a functor introduced by Cochran and Habegger [CH]. For g in $\mathbf{UF}(odd)$ we choose a map

$$w_2^g : X^g \to K(\mathbb{Z}/2, 2) = K \tag{1}$$

representing the Stiefel-Whitney class. Here X^g is a Poincaré space representing g. We may assume that w_2^g is a fibration. One can check that homotopy classes in $[X^f, X^g]_{\xi,\eta}$ of pointed maps and homotopy classes in ${}_K[X^f, X^g]_{\xi,\eta}$ of pointed maps over K coincide, i.e.

$$[X^f, X^g]_{\xi,\eta} \;=\; {}_K[X^f, X^g]_{\xi,\eta} \tag{2}$$

for all $(\xi, \eta) : f \to g$ in $\mathbf{UF}(odd)$. For maps $F \in {}_K[X^f, X^g]_{\xi,\eta}$ Cochran and Habegger (4.3 in [CH]) use a functor Ω on the category of spaces X over $K(\mathbb{Z}/2, 2)$. This functor is defined as follows. Consider the following pull back diagrams:

$$\begin{array}{ccc}
\bar{\bar{X}} & \longrightarrow & ESO(n) \\
\downarrow{\scriptstyle \bar{f}*\gamma_n} & & \downarrow{\scriptstyle \gamma_n} \\
\bar{X} & \xrightarrow{\bar{f}} & BSO(n) \\
\downarrow & & \downarrow{\scriptstyle \psi} \\
X & \xrightarrow[f]{} & K(\mathbb{Z}/2, 2)
\end{array}$$

Here γ_n is the universal n-plane bundle over $BSO(n)$ and ψ is the fibration with fiber $BSpin(n)$. Let $T_n(X, f)$ be the Thom space of the n-plane bundle $\bar{f}^*\gamma_n$ and let

$$h_*(X, f) = \lim_{n\to\infty} \pi_* T_n(X, f).$$

Then h_* is a homology theory for spaces over $K(\mathbb{Z}/2, 2)$. For a basepoint $x_0 \in X$ the group $\Omega(X, f)$ is defined by the relative group

$$\Omega(X, f) \;=\; h_4(X, x_0, f)$$

denoted by $\tilde{\Omega}_4^{w_2}(X)$ in [CH]. Cochran and Habegger [CH] (using a result of Rochlin [R]) obtain for the (2,4)-Poincaré-complex X^g the following natural commutative diagram with exact rows.

$$(3)\qquad \begin{array}{ccccccccc} 0 & \longrightarrow & H_2(X^g, \mathbb{Z}/2) & \longrightarrow & \Omega(X^g) & \longrightarrow & H_4(X^g) & \longrightarrow & 0 \\ & & \downarrow{\scriptstyle w_2^g} & & \downarrow{\scriptstyle \alpha^g} & & \downarrow{\scriptstyle \beta^g} & & \\ 0 & \longrightarrow & \mathbb{Z}/2 & \longrightarrow & \mathbb{Z}/16 & \longrightarrow & \mathbb{Z}/8 & \longrightarrow & 0 \end{array}$$

The map F in ${}_K[X^f, X^g]_{\xi,\eta}$ induces a map of the top row over the bottom row, that is $\alpha^g\Omega(F) = \alpha^f$ and $\beta^g\xi = \beta^g H_4(F) = \beta^f$. According to the remark following 4.4 in [CH] Cochran and Habegger do not identify β^g in terms of g. We check in (2.4.9) that β^g, in fact, coincides with the signature τ^g. This shows that diagram (3) is an object as in $\mathbf{S}(odd)$. Using the proof of 4.5 in [CH] we see that the functor

$$(4)\qquad \Omega : \mathbf{P}(odd) \to \mathbf{S}(odd)$$

given by the functor Ω above is well defined. In fact Ω satisfies by 4.5 in [CH] the formula $\Omega(F+\alpha) = \Omega(F)+\alpha'$ where α' is the K^u coordinate of $\alpha \in D(\xi, \eta)$. Here we use the fact that $L(\xi, \eta)^2 = 0$. Hence Ω is an equivalence of linear extensions. □

2.4.9. LEMMA: *The Cochran and Habegger map β^g is the signature τ^g.*

Proof: For g we have the mapping cone C_g of $g : S^3 \to M(A, 2)$. A characteristic element v^g of the form g in (2.4.4)(1) yields a push out diagram

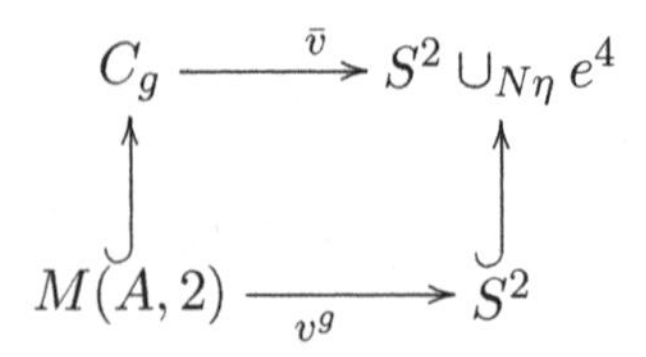

where $N = \Gamma(v^g)(g(1)) = v^g \cup v^g \equiv \tau_g \; mod 8$. Now we observe that the map (2.4.8)(1) can be obtained by a factorization

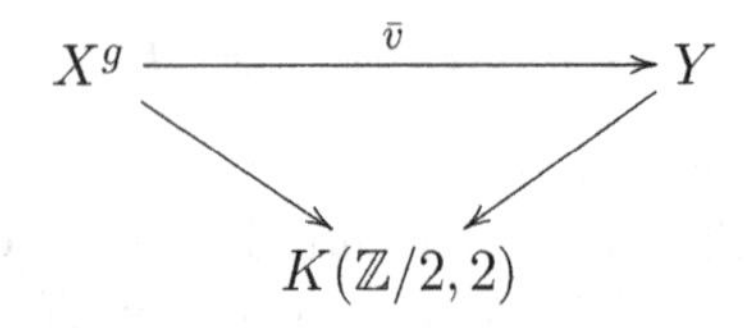

where $Y \simeq S^2 \cup_{N\eta} e^4$. The computation of $\Omega(Y)$ now shows that $\beta^g = \tau^g$, see [CH]. For this we use the fact that $\beta : H^4 X \to \mathbb{Z}/8$ is surjective if X is the 4-skeleton of $K(\mathbb{Z}/2, 2)$ given by $S^2 \cup_2 e^3 \cup_\eta e^4$. □

2.4.10. Definition: Consider the short exact sequence

$$0 \longrightarrow \mathbb{Z}/2 \xrightarrow{\;i\;} \mathbb{Z}/16 \xrightarrow{\;p\;} \mathbb{Z}/8 \longrightarrow 0\cdot$$

For $(\xi, \eta) : f \to g$ in $\mathbf{UF}(odd)$ we know by (2.4.4) that $\tau_f \equiv \xi \tau_g \; mod 8$. Hence the *signature derivation* Δ with

$$\Delta(\xi, \eta) \;=\; i^{-1}(\tau_f - \xi\tau_g) \;\in \mathbb{Z}/2$$

is well defined. One readily checks the composition formula

$$\Delta((\xi', \eta')(\xi, \eta)) \;=\; \Delta(\xi', \eta') + \Delta(\xi, \eta)$$

where ξ' and ξ are odd.

Let $\mathbf{U} \subset \mathbf{UF}(odd)$ be a subcategory. We say that $\mathbf{P}(odd)$ is *split over* $\mathbf{U}$ if there is a functor s for which the following diagram commutes:

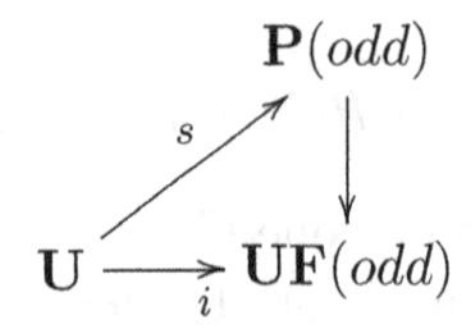

where i is the inclusion. As a consequence of (2.4.8) we get the following result.

2.4.11. THEOREM: *$\mathbf{P}(odd)$ is split over $\mathbf{U}$ with $\mathbf{U} \subset \mathbf{UF}(odd)$ if and only if there exists a derivation δ on $\mathbf{U}$ with*

$$\delta(\xi, \eta) \in A \otimes \mathbb{Z}/2$$

for $(\xi, \eta) : f \to g$ in $\mathbf{U}$, $g : \mathbb{Z} \to \Gamma(A)$, satisfying

$$\delta((\xi, \eta)(\xi', \eta')) = (\eta \otimes \mathbb{Z}/2)\delta(\xi', \eta') + \delta(\xi, \eta)$$

for $(\xi', \eta') : h \to f$ in $\mathbf{U}$ such that δ is a lift of the signature derivation Δ, that is

$$w_2^g \delta(\xi, \eta) \;\; = \;\; \Delta(\xi, \eta)$$

for (ξ, η) in $\mathbf{U}$.

Proof: We consider the category $\mathbf{S}(odd)$ in (2.4.6) where we choose for each g a splitting

$$\Omega(g) = (A \otimes \mathbb{Z}/2) \oplus \mathbb{Z}$$

compatible with the top row of (2.4.6)(1). Then a splitting s is given by

$$s(\xi, \eta) = (\xi, \eta, \lambda(\xi, \eta))$$

with $\lambda(\xi, \eta) : \Omega(f) \to \Omega(g)$ given by a matrix

$$\lambda(\xi, \eta) \;\; = \;\; \begin{pmatrix} \eta \otimes \mathbb{Z}/2 & , & \delta(\xi, \eta) \\ 0 & , & \xi \end{pmatrix}$$

and $\delta(\xi, \eta) \in A \otimes \mathbb{Z}/2$. Now the derivation property for δ corresponds exactly to the formula $s((\xi, \eta)(\xi', \eta')) = s(\xi, \eta) \circ s(\xi', \eta')$. Moreover one readily checks by the definition of $\mathbf{S}(odd)$ that δ is a lift of Δ. □

We say that the *signature obstruction vanishes* over $\mathbf{U}$ if a lift δ of Δ exists over $\mathbf{U}$ as in (2.4.11). We can reformulate this as follows. The signature τ

yields an element

$$\tau \in H^0(\mathbf{UF}(odd), \mathbb{Z}/8).$$

Using the boundary

$$\partial : H^0(\mathbf{UF}(odd), \mathbb{Z}/8) \to H^1(\mathbf{UF}(odd), \mathbb{Z}/2)$$

induced by the short exact sequence $0 \to \mathbb{Z}/2 \to \mathbb{Z}/16 \to \mathbb{Z}/8 \to 0$ we obtain the element

$$\partial\tau = \Delta \in H^1(\mathbf{UF}(odd), \mathbb{Z}/2) \tag{2.4.12}$$

represented by the signature derivation. On the other hand we have a map of natural systems

$$w : R(\xi, \eta) = A \otimes \mathbb{Z}/2 \twoheadrightarrow \mathbb{Z}/2$$

for $(\xi, \eta) : f \to g$ in $\mathbf{UF}$ with $w = w_2^g$. The kernel of w is the natural system K^u so that we have a short exact sequence of natural systems on $\mathbf{UF}(odd)$

$$0 \to K^u \to R \to \mathbb{Z}/2 \to 0.$$

This induces the exact sequence for $i : \mathbf{U} \subset \mathbf{UF}(odd)$:

$$H^1(\mathbf{U}, R) \xrightarrow{w_*} H^1(\mathbf{U}, \mathbb{Z}/2) \xrightarrow{\partial'} H^2(\mathbf{U}, K^u).$$

2.4.13. Corollary: *The signature obstruction over* $\mathbf{U}$ *is the element*

$$\partial' i^* \Delta \in H^2(\mathbf{U}, K^u)$$

where $\Delta = \partial\tau$ *above. The signature obstruction over* $\mathbf{U}$ *vanishes, i.e.* $\partial' i^* \Delta = 0$*, if and only if there is a lift* δ *of* Δ *as in (2.4.11) or equivalently if and only if* $\mathbf{P}(odd)$ *splits over* $\mathbf{U}$.

Of course $\partial' i^* \Delta$ coincides with the element represented by the linear extension $K^u \to \mathbf{P}(odd)|_{\mathbf{U}} \to \mathbf{U}$ which is a restriction of the linear extension (2.4.3). In particular we get for $\mathbf{U} = \mathbf{UF}(odd)$

$$(2.4.14) \qquad <\mathbf{P}(2,4)_0> = \partial'\partial(\tau) \in H^2(\mathbf{UF}_0, K^u) = H^2(\mathbf{UF}(odd), K^u)$$

so that the coordinate $<\mathbf{P}(2,4)_0>$ in (2.3.1) is computed in terms of the signature τ.

2.4.15. COROLLARY: *Assume all objects g in $\mathbf{U}$ have a signature τ_g with $\tau_g \equiv 0\ mod 8$. Then $\mathbf{P}(odd)$ splits over $\mathbf{U}$.*

2.4.16. COROLLARY: *Assume all objects $g : \mathbb{Z} \to \Gamma(A)$ in $\mathbf{U}$ satisfy $rank(A)$ is odd. Then $\mathbf{P}(odd)$ splits over $\mathbf{U}$.*

Proof: If $rank(A) = odd$ we have the element

$$\sigma(g(1)) \in A \otimes \mathbb{Z}/2 \xrightarrow{w_2^g} \mathbb{Z}/2$$

with $w_2^g(\sigma(g(1)) = 1$, see 3.5 in [HNK]. Hence $w : R\ |_{\mathbf{U}} \to \mathbb{Z}/2$ admits a splitting of natural systems so that ∂' in (2.4.13) is trivial.

□

2.5. The group of homotopy equivalences

Let X be a simply connected 4-manifold or a (2,4)-Poincaré complex. The group of homotopy equivalences $Aut(X)$ is the group of invertible elements in the monoid $[X, X]$. If $g : \mathbb{Z} \to \Gamma(A)$ is the form of X then $Aut(g)$ is the group of $\pm$-automorphisms of g, i.e. the group of all $\varphi \in Aut(A)$ with $\Gamma(\varphi)g = \pm g$. By (1.3.12) we have the following group extension which was already obtained by Baues in [BW].

$$(2.5.1) \qquad 0 \to ker(w_2^g \otimes \mathbb{Z}/2) \to Aut(X) \to Aut(g) \to 0.$$

Here we use the computation of the natural system D in (2.2.10) For $\varphi \in Aut(g)$ we obtain $ker(w_2^g \otimes \mathbb{Z}/2 : A \otimes \mathbb{Z}/2 \to \mathbb{Z}/2)$ as a right $Aut(g)$-module by $x^\varphi = (\varphi^{-1} \otimes \mathbb{Z}/2)(x)$.

2.5.2. THEOREM: *The extension $Aut(X) \to Aut(g)$ is split.*

Proof: If $w_2^g = 0$ this is a consequence of (2.3.10). If $w_2^g \neq 0$ and if there is no element in $Aut(X)$ of degree -1 then this follows from (2.4.11). If there is an element in $Aut(X)$ od degree -1 then the lemma below shows that $\tau(g) = 0$ and hence the theorem follows from (2.4.14). □

2.5.3. LEMMA: *Let $g : \mathbb{Z} \to \Gamma(A)$ be an unimodular form for which there exists an automorphiism $\varphi \in Aut(A)$ with $\Gamma(\varphi)g = -g$. Then the signature τ_g is trivial.*

Proof: The from g has to be indefinite and hence g splits for $A = A_1 \oplus A_2$.

$$g = \begin{pmatrix} g_1 & 0 \\ 0 & g_2 \end{pmatrix}$$

where g_1 is positive definite and g_2 is negative definite. Now $\varphi A_1 \oplus \varphi A_2$ is again a splitting of the form g by 1.4 of [HNK]. Hence $rank(A_1) \leq rank(A_2)$ and $rank(A_2) \leq rank(A_1)$. Therefore the signature $\tau_g = rank(A_1) - rank(A_2) = 0$. □

2.5.4. REMARK: Cochran and Habegger [CH] study the group of homotopy equivalences of a 1-connected 4-manifold X in great detail and our proof of (2.5.1) above relies on the method in [CH] which in turn uses at a crucial point a result of Rochlin [R]. There is however a gap in the proof of [CH] since lemma 4.4 of [CH] does not hold; here an obstruction arises as in the proof of (2.4.11) if $Aut(\mathbb{Z} \to \mathbb{Z}/2) \neq 0$. In this case we need the identification of β^g and τ_g in (2.4.9) which is not contained in [CH]. Having this identification we use lemma (2.5.3) which thus yields the splitting. We refer the reader to [CH] for many further properties of the group extension (2.5.1). Also in [CH] one finds a list of mistakes in the literature ([KS], [M], [Q], [WH]) concerning the group $Aut(X)$.

2.5.5. REMARK: Let $Aut^+(X) \subset Aut(X)$ and $Iso(g) \subset Aut(g)$ be the subgroups of orientation preserving automorphisms. Then a splitting of the extension

$$0 \to ker(w_2^g \otimes \mathbb{Z}/2) \to Aut^+(X) \to Iso(g) \to 0$$

can also be obtained geometrically as follows. If X is a 1-connected 4-manifold then Quinn [Q] shows that the group $Iso(X)$ of isotopy classes of orientation preserving homeomorphisms $X \to X$ is isomorphic to $Iso(g)$. Hence the canonical map $Iso(X) \to Aut^+(X)$ is a splitting. On the other hand Kreck ([K], [KH]) computes the diffeomorphism classes relative to the boundary of orientated h-cobordisms between X and X. This also gives a splitting of $Aut^+(X) \to Iso(G)$.

2.6. The Lie part

We define a natural equivalence relation $\sim$ on the category $\mathbf{P}(2,4)$ of 1-connected 4-dimensional Poincaré complexes. For maps $F, G : C_f \to C_g$ in $\mathbf{P}(2,4)$ we write $F \sim G$ if F and G induce the same morphism in homology, i.e. $H_*(F) = H_*(G) = (\xi, \eta) : f \to g$ in $\mathbf{UF}$ and if there is $\alpha \in \pi_3 C_g$ such that $F + \alpha(\Sigma\eta) = G$ where $\Sigma\eta : S^4 \to S^3$ is the suspension of the Hof map. This is a natural equivalence relation which yields the quotient category

$$\mathbf{P}^L(2,4) = \mathbf{P}(2,4)/\sim = \mathbf{P}(2,4)/D^\Gamma.$$

According to (2.2.8) we obtain the linear extension of categories

$$L \to \mathbf{P}^L(2,4) \to \mathbf{UF} \tag{2.6.1}$$

which we call the Lie part of $\mathbf{P}(2,4)$. Here L is the natural system (2.2.7). In this section we describe an algebraic category $\mathbf{L}(2,4)$ equivalent to $\mathbf{P}^L(2,4)$.

A group G has nilpotency degree 2 or is a nil(2)-group if all triple commutators vanish in G. For a finitely generated free abelian group A we choose a *free nil(2)-group* E_A with abelianization

$$E_A^{ab} = A.$$

For example, if $e_1, \ldots, e_n$ are a basis of A then we can choose $E_A = < e_1, \ldots, e_n > /\Gamma_3$ where Γ_3 is the subgroup of the free group $< e_1, \ldots, e_n >$ generated by all triple commutators. The *exterior square* $\Lambda^2 A$ is given by the natural short exact sequence

$$0 \longrightarrow \Gamma(A) \xrightarrow{\tau} A \otimes A \xrightarrow{q} \Lambda^2 A \longrightarrow 0.$$

For $a, b \in A$ we write $a \wedge b = q(a \otimes b) \in A \wedge A = \Lambda^2 A$. There is the well known central extension of groups

$$0 \longrightarrow \Lambda^2 A \xrightarrow{w} E_A \xrightarrow{ab} A \longrightarrow 0 \tag{2.6.2}$$

where ab is the abelianization and w is the commutator map. We write the group structure of E_A additively so that for $x, y \in E_A$ the commutator $(x, y) = -x - y + x + y$ needs not to be trivial. Let $\{x\} \in A$ be the element represented by $x \in E_A$; then $w(\{x\} \wedge \{y\}) = (x, y)$.

Let **ab** be the category of finitely generated abelian groups and let **nil** be the category of groups E_A with A in **ab**. Then we have a linear extension of categories

$$(2.6.3) \qquad Hom(\_, \Lambda^2) \longrightarrow \mathbf{nil} \xrightarrow{ab} \mathbf{ab}$$

where the abelianization functor ab carries E_A to A. Given $\eta, \eta' : E_{A'} \to E_A$ in **nil** with $ab(\eta) = ab(\eta') = \bar{\eta} : A' \to A$ we obtain for η (η') the commutative diagram

$$\begin{array}{ccc} A' \wedge A' & \xrightarrow{\bar{\eta}\wedge\bar{\eta}} & A \wedge A \\ \downarrow & & \downarrow w \\ E_{A'} & \xrightarrow{\eta\ (\eta')} & E_A \\ ab \downarrow & & \downarrow \\ A' & \xrightarrow{\bar{\eta}} & A \end{array}$$

so that there exists a unique element $\alpha \in Hom(A', \Lambda^2 A)$ with $\eta' = \eta + w\alpha ab = \eta + \alpha$. We thus define the natural system $Hom(\_, \Lambda^2)$ on **ab** by $Hom(\_, \Lambda^2)(\bar{\eta}) = Hom(A', \Lambda^2 A)$. In [BD] we show

2.6.4. THEOREM: *The cohomology class* $<\mathbf{nil}>$ *represented by (2.6.3) is the generator in*

$$<\mathbf{nil}> \in H^2(\mathbf{ab}, Hom(\_, \Lambda^2)) = \mathbb{Z}/2.$$

We need the following morphism of natural systems.

2.6.5. DEFINITION: For $f : \mathbb{Z} \to \Gamma(A')$ and $g : \mathbb{Z} \to \Gamma(A)$ and $(\xi, \eta) : f \to g$ in **UF** let

$$(1) \qquad \psi : Hom(A', \Lambda^2 A) \to L(\xi, \eta)$$

be defined as follows; here $L(\xi, \eta)$ is defined by the push out in (2.3.7). For

$\alpha : A' \to \Lambda^2 A$ we choose $H : A' \to A \otimes A$ with $qH = \alpha$. Then we obtain the composite

$$\mathbb{Z} \xrightarrow{f} \Gamma(A') \xrightarrow{\tau} A \otimes A \xrightarrow{\eta\otimes H - H\otimes\eta} A \otimes A \otimes A \xrightarrow{[[1,1],1]} L(A,1)_3 \twoheadrightarrow L(\xi,\eta)$$

which carries 1 to $-\psi(\alpha)$. The following lemma shows that $\psi(\alpha)$ does not depend on the choice of H. If we consider $Hom(\_, \Lambda^2)$ as a natural system on **UF** via the forgetful functor **UF** $\to$ **ab** then ψ is a homomorphism of natural systems

$$\psi : Hom(\_, \Lambda^2) \to L \tag{2}$$

2.6.6. LEMMA: *The element $\psi(\alpha)$ does not depend on the choice of H.*

Proof: Let $\beta : A' \to \Gamma(A)$ and consider $H' = H + \tau\beta$. Then the result follows since the composite

$$l(\beta) : \; A' \otimes A' \xrightarrow{\eta\otimes\tau\beta - \tau\beta\otimes\eta} A \otimes A \otimes A \twoheadrightarrow L(\xi,\eta)$$

is trivial. The group $Hom(A', \Gamma(A))$ is generated by composites

$$A' \xrightarrow{n} \mathbb{Z} \longrightarrow \Gamma(A)$$

where $\mathbb{Z} \to \Gamma(A)$ carries 1 to $\gamma(a)$ with $a \in A$. Since $l(\beta)$ is linear in β we thus may assume that β is such a composite. Hence $l(\beta)$ carries $x \otimes y$ to $n(y)[[\eta x, a], a] + n(x)[[a,a], \eta y]$ where $\bar{y} = \eta y$ satisfies

$$\begin{aligned} [[a,a],\bar{y}] &= -[[\bar{y},a],a] - [[a,\bar{y},a] \\ &= -2[[\eta y, a], a]. \end{aligned}$$

By definition of $L(\xi,\eta)$ in (2.2.7) the elements $[[\eta y, a], a]$ and $[[\eta y, a], a]$ represent zero in $L(\xi,\eta)$. $\square$

2.6.7. DEFINITION: The category $\mathbf{L}(2,4)$ is defined as a linear extension of **UF** by the following diagram:

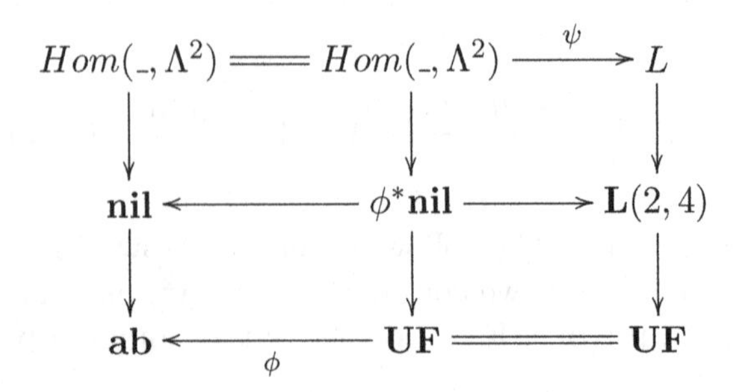

Here we use a pull back and a push out of linear extensions. More explicitly the category $\mathbf{L}(2,4)$ can be defined as follows. Objects are the same as in **UF** and morphisms $f \to g$ with $f : \mathbb{Z} \to \Gamma(A')$, $g : \mathbb{Z} \to \Gamma(A)$ are equivalence classes $\{(\xi, \eta, \delta)\}$ of triples (ξ, η, δ) with $\xi : \mathbb{Z} \to \mathbb{Z}$, $\eta : E_{A'} \to E_A$ such that $(\xi, \eta^{ab}) : f \to g$ is a morphism in **UF** and $\delta \in L(\xi, \eta^{ab})$. The equivalence relation is given by $(\xi, \eta + \alpha, \delta) \sim (\xi, \eta, \psi(\alpha) + \delta)$, for $\alpha \in Hom(A', \Lambda^2 A)$; see (2.6.3).

2.6.8. THEOREM: *There is an isomorphism of linear extensions*

$$\mathbf{P}^L(2,4) \cong \mathbf{L}(2,4).$$

The theorem is a corollary of (4.5.6) below.

2.6.9. COROLLARY: *The cohomology class* $< \mathbf{P}^L(2,4) >$ *represented by the Lie part (2.6.1) satisfies*

$$< \mathbf{P}(2,4) >= \psi_* \phi^* < \mathbf{nil} > .$$

Here we use the composite

$$\mathbb{Z}/2 = H^2(\mathbf{ab}, Hom(\_, \Lambda^2)) \xrightarrow{\phi^*} H^2(\mathbf{UF}, Hom(\_, \Lambda^2)) \xrightarrow{\psi_*} H^2(\mathbf{UF}, L)$$

where $\phi : \mathbf{UF} \to \mathbf{ab}$ *is the forgetful functor and where* ψ *is the map between natural systems in (2.6.5).*

2.6.10. COROLLARY: *For odd primes the coordinate*

$$0 =< \mathbf{P}(2,4)_0 >^p \in H^2(\mathbf{UF}_0, D^p)$$

in (2.2.6) is trivial.

Proof (of (2.6.9)): For any finite subcategory $\mathbf{U} \subset \mathbf{UF}_0$ there is a prime power p^n with $p^n(D^p\ |_{\mathbf{U}}) = 0$. Hence $p^n < \mathbf{P}(2,4)_0 >^p|_{\mathbf{U}} = 0$ so that $< \mathbf{P}(2,4)_0 >^p|_{\mathbf{U}} = 0$ since by (2.2.10) the p-coordinate $< \mathbf{P}^L(2,4)_0 >^p$ coincides with $< \mathbf{P}(2,4)_0 >^p$ and has order 2 by (2.6.8). □

2.7. The homotopy category of maps with non-trivial degree

Recall that the homotopy category $\mathbf{P}(2,4)_0$ of maps in $\mathbf{P}(2,4)$ with non-trivial degree is part of a linear extension

$$D \to \mathbf{P}(2,4)_0 \to \mathbf{UF} \tag{2.7.1}$$

where D splits as $D = L \oplus K^{st} \oplus K^u$ and L splits as a product $L = \times_{p \text{ prime}} L^p$. Hence the cohomology class $< \mathbf{P}(2,4)_0 >$ is determined by the coordinates

$$\begin{aligned} < \mathbf{P}^L(2,4)_0 >^p &\in H^2(\mathbf{UF}_0, L^p)\ ,\ p \text{ prime}, \\ < \mathbf{P}^{st}(2,4)_0 > &\in H^2(\mathbf{UF}_0, K^{st}), \\ < \mathbf{P}^u(2,4)_0 > &\in H^2(\mathbf{UF}_0, K^u). \end{aligned}$$

These coordinates are completely algebraically determined in the next result. Let $q_2 : L \to L^2$ be the projection.

2.7.2. THEOREM: *We have $< \mathbf{P}^L(2,4)_0 >^p = 0$ for p odd. Moreover we have*

$$\begin{aligned} < \mathbf{P}^L(2,4)_0 > &= (q_2\psi)_*\phi^* < \mathbf{nil} > \ (see (2.6.8)), \\ < \mathbf{P}^{st}(2,4)_0 > &= < \mathbf{S}(2,4)_0 > \ (see (2.3.9)), \\ < \mathbf{P}^u(2,4)_0 > &= \partial'\partial(\tau)\ (see (2.4.14)). \end{aligned}$$

This result can be deduced and in fact corresponds to the following algebraic computation of the category $\mathbf{P}(2,4)_0$.

2.7.3. THEOREM: *The category $\mathbf{P}(2,4)_0$ as a linear extension is equivalent to the pull back of linear extensions*

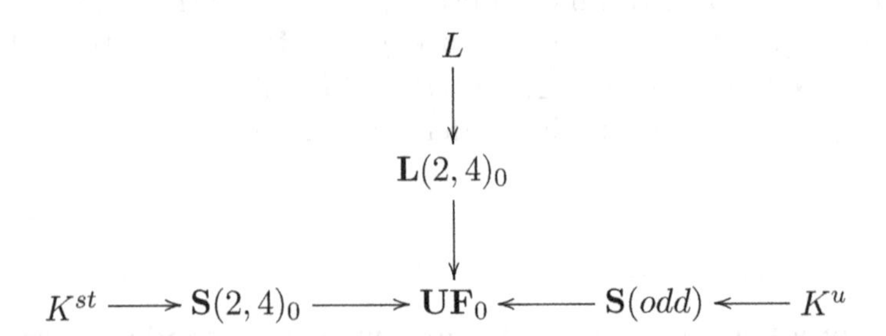

Compare (2.3.8), (2.4.6), (2.6.7). More precisely the pull back category $\bar{\mathbf{P}}(2,4)_0$ *has objects* f, g *as in* $\mathbf{UF}_0$ *and morphisms* $f \to g$ *are triples* (A, B, C) *where* $A : f \to g$, $B : f \to g$, $C : f \to g$ *are morphisms in* $\mathbf{S}(2,4)_0$, $\mathbf{L}(2,4)_0$ *and* $\mathbf{S}(odd)$ *respectively such that* A, B *and* C *induce the same morphism* $(\xi, \eta) : f \to g$ *in* $\mathbf{UF}_0$. *There is an equivalence of linear extensions*

$$\bar{\mathbf{P}}(2,4)_0 = \mathbf{P}(2,4)_0$$

compatible with the isomorphism $K^{st} \oplus L \oplus K^u = D$.

For each 1-connected 4-manifold X with intersection form $g : \mathbb{Z} \to \Gamma(A)$, $A = H_2X$, we have the following commutative diagram of homomorphisms between monoids:

(2.7.4)
$$\begin{array}{ccccccc} Aut(X) & \hookrightarrow & [X,X]_{odd} & \hookrightarrow & [X,X]_0 & \hookrightarrow & [X,X] \\ \downarrow p_{Aut} & & \downarrow p_{odd} & & \downarrow p_0 & & \downarrow p \\ Aut(g) & \hookrightarrow & End(g)_{odd} & \hookrightarrow & End(g)_0 & \hookrightarrow & End(g) \end{array}$$

Here $End(g)$ is the monoid of endomorphisms of g in $\mathbf{UF}$. The subscript "*odd*" and the 0 indicate the submonoids of maps of odd degree and non-trivial degree respectively. Theorem (2.5.1) shows that there is always a homomorphism $q_{Aut} : Aut(g) \to Aut(X)$ of monoids which is a splitting of p_{Aut}. Does such a splitting also exist for p_{odd}, p_0 and p? Concerning this problem we get the following result where we use the signature obstruction in (2.4.11).

2.7.5. THEOREM: *A homomorphism*

$$q_{odd} : End(g)_{odd} \to [X,X]_{odd}$$

of monoids which is a splitting of p_{odd} *(i.e.* $p_{odd}q_{odd} = 1$*) exists if and only if the signature obstruction for* $\mathbf{U} = End(g)_{odd}$ *is trivial.*

Proof: We use the fact that L^2 is trivial on $End(g)_{odd}$ and we use (2.3.10). Hence the only remaining obstruction is the signature obstruction (2.4.11). □

2.7.6. COROLLARY: *A splitting q_{odd} exists if one of the following conditions is satisfied:*

- *Stiefel-Whitney class w_2^g of X is trivial,*
- *$rank(H_2X)$ is odd,*
- *signature of X is congruent to 0 $mod8$.*

Otherwise the signature obstruction might be non-trivial if $w_2^g \neq 0$, $rank(H_2X)$ is even and $signature(X) \not\equiv 0 \ mod8$. The example (2.7.8) below shows the kind of computation needed to identify the signature obstruction.

2.7.7. THEOREM: *Assume a splitting q_{odd} of p_{odd} exists. Then a splitting homomorphism*

$$q_0 : End(g)_0 \to [X, X]_0$$

of p_0 exists if and only if the Lie obstruction (see(2.6.8))

$$(q_2\psi)_* < \mathbf{nil} > |_{End(g)_0} = 0$$

vanishes. Here $q_2 : L \to L^2$ is the projection of L onto the component of L for the prime 2.

Proof: This is a consequence of (2.7.2) since the existence of p_{odd} implies that the signature obstruction vanishes. Moreover we use (2.3.10). □

We now describe an example for which the signature obstruction does not vanish.

2.7.8. THEOREM: *The signature obstruction for the connected sum*

$$X = \mathbb{C}P_2 \# \mathbb{C}P_2$$

is non trivial. Hence for the form g of X a splitting homomorphism q_{odd} of the surjection

$$p_{odd} : [X, X]_{odd} \twoheadrightarrow End(g)_{odd}$$

does not exist.

Proof: Let e_1, e_2 be the canonical basis of $A = H_2X = \mathbb{Z} \oplus \mathbb{Z}$ so that $g(1) = \gamma(e_1) + \gamma(e_2) \in \Gamma(A)$. For

$$(1) \qquad \varphi = \begin{pmatrix} a & b \\ c & d \end{pmatrix}$$

we have $\varphi \in End(g)$ if and only if there is ξ with $\Gamma(\varphi)(\gamma(e_1) + \gamma(e_2)) = \xi(\gamma(e_1) + \gamma(e_2))$ where

$$\begin{aligned} & \Gamma(\varphi)(\gamma(e_1) + \gamma(e_2)) = \gamma(ae_1 + ce_2) + \gamma(be_1 + de_2) \\ = \quad & a^2\gamma(e_1) + ac[e_1, e_2] + c^2\gamma(e_2) + b^2\gamma(e_1) + bd[e_1, e_2] + d^2\gamma(e_2). \end{aligned}$$

Hence we get $a^2 + b^2 = \xi$, $c^2 + d^2 = \xi$ and $ac = -bd$. Thus $a^2c^2 = b^2d^2 = (\xi - a^2)(\xi - c^2)$ and hence $0 = \xi(\xi - a^2 - c^2)$. Therefore $\xi = 0$ or $\xi = a^2 + c^2$ if $\xi \neq 0$. Hence if $\xi \neq 0$ we get $b^2 = c^2$, $d^2 = a^2$. Therefore we see that $End(g)_0$ is generated as a monoid by the matrices

$$(2) \qquad E = \begin{pmatrix} 1 & 0 \\ 0 & -1 \end{pmatrix} \text{ and } F = \begin{pmatrix} a & b \\ -b & a \end{pmatrix} = a + bi$$

with $a, b \in \mathbb{Z}$ and $a^2 + b^2 \neq 0$. We have the composition formula $(a + bi) \cdot (a' + b'i) = aa' - bb' + (ab' + ba')i$ in the same way as in complex numbers $\mathbb{C}$. Let $M \subset End(g)_{odd}$ be the submonoid consisting of matrices E with $a^2 + b^2 = \xi$ odd. Consider the commutative diagram

$$(3) \qquad \begin{array}{ccc} \mathbb{Z}/2 \oplus \mathbb{Z}/2 & = & A \otimes \mathbb{Z}/2 \\ \downarrow{\scriptstyle w} & & \downarrow{\scriptstyle \bar{w}} \\ \mathbb{Z}/2 & \xrightarrow{i} & \mathbb{Z}/16 \end{array}$$

with $w = w_2^g$ given by $w(e_1) = w(e_2) = 1$. According to (2.4.11) the signature obstruction vanishes on $M = \mathbf{U}$ if and only if there is $\delta(F) \in A \otimes \mathbb{Z}/2$ for $F \in M$ with

$$(4) \qquad \delta(FF') = (F \otimes \mathbb{Z}/2)\delta(F') + \delta(F), \text{ and}$$

$$(5) \qquad iw\delta(F) = \tau_g - \xi\tau_g \in \mathbb{Z}/16$$

where $\tau_g = 2$. Since $\xi = a^2 + b^2$ is odd we see that for $a = 2n$, $b = 2m + 1$ we have

$$\begin{aligned} \tau_g - \xi\tau_g &= 2 - 2(4n^2 + (2m+1)^2) \\ &\equiv 8(n^2 + m^2 + m) \\ &\equiv 8n \; mod 16 \end{aligned}$$

Hence we see that (5) is equivalent to

$$w\delta(a + bi) = \{a \cdot b/2\} \in \mathbb{Z}/2 \tag{6}$$

where $a \cdot b/2 \in \mathbb{Z}$ is well defined since $a^2 + b^2$ is odd. We now write

$$\delta(F) = \delta_1(F)e_1 \oplus \delta_2(F)(e_1 + e_2) \in A \otimes \mathbb{Z}/2 \tag{7}$$

with $\delta_1(F), \delta_2(F) \in \mathbb{Z}/2$. Then (6) shows $\delta(F) = \{a \cdot b/2\}$. Moreover (4) is equivalent to the following equations where $\delta_1' = \delta_1(F'), \delta_2' = \delta_2(F')$.

$$\begin{aligned} \delta(FF') &= \delta(FF')e_1 + \delta_2(FF')(e_1 + e_2) \\ &= (F \otimes \mathbb{Z}/2)\delta(F') + \delta(F) \\ &= (F \otimes \mathbb{Z}/2)(\delta_1' e_1 + \delta_2'(e_1 + e_2)) + \delta_1 e_1 + \delta_2(e_1 + e_2) \\ &= \delta_1'(ae_1 - be_2) + \delta_2'(e_1 + e_2) + \delta_1 e_1 + \delta_2(e_1 + e_2) \\ &= (\delta_1'(a + b) + \delta_1)e_1 + (\delta_2 + \delta_2' - \delta_1' b)(e_1 + e_2) \\ &= (\delta_1' + \delta_1)e_1 + (\delta_2 + \delta_2' + \delta_1' b)(e_1 + e_2). \end{aligned}$$

Hence (4) is equivalent to

$$\delta_2(FF') = \delta_2(F) + \delta_2(F') + \{a' \cdot b'/2\} \cdot b \; \in \mathbb{Z}/2 \tag{8}$$

where $F' = a' + b'i$ and $\delta_1' = \delta_1(F') = \{a' \cdot b'/2\}$. We now apply (8) to the products

$$(a + bi) \cdot i = -b + ai = i \cdot (a + bi)$$

Hence we get

$$\begin{array}{rclr} \delta_2((a+bi)\cdot i) & = & \delta_2(a+bi)+\delta_2(i)+0 & = \\ \delta_2(i\cdot(a+bi)) & = & \delta(i)+\delta_2(a+bi)+\{a\cdot b/2\} & \end{array}$$

For $a = 1$, $b = 2$ this is a contradiction. Hence δ with the properties (4), (5) does not exist and therefore the signature obstruction does not vanish. □

2.7.9. EXAMPLE: The proof of (2.7.8) shows the following fact. There exist maps

$$F_{a,b}, T : \mathbb{C}P_2\#\mathbb{C}P_2 \to \mathbb{C}P_2\#\mathbb{C}P_2$$

which induce in $H_2(\mathbb{C}P_2\#\mathbb{C}P_2) = \mathbb{Z}\oplus\mathbb{Z}$ the matrices

$$H_2(F_{a,b}) = \begin{pmatrix} a & b \\ -b & a \end{pmatrix}, \quad H_2(T) = \begin{pmatrix} 0 & 1 \\ -1 & 0 \end{pmatrix}$$

with $a, b \in \mathbb{Z}$ and $a + b$ odd. We have

$$\begin{pmatrix} a & b \\ -b & a \end{pmatrix}\begin{pmatrix} 0 & 1 \\ -1 & 0 \end{pmatrix} = \begin{pmatrix} -b & a \\ -a & -b \end{pmatrix} = \begin{pmatrix} 0 & 1 \\ -1 & 0 \end{pmatrix}\begin{pmatrix} a & b \\ -b & a \end{pmatrix}$$

in homology. Moreover if $a\cdot b/2$ is even then $F_{a,b}$ and T can be chosen such that

$$F_{a,b}T \simeq TF_{ab}$$

is a homotopy. However if $a \cdot b/2$ is odd then all choices $F_{a,b}$ and T satisfy

$$F_{a,b}T \not\simeq TF_{ab}.$$

CHAPTER 3

TRACK CATEGORIES

We show that the homotopy category $\mathbf{CW}(2,4)$ of CW-complexes with cells only in dimensions 2 and 4 is equivalent to the category $\mathbf{T}(2,4)/\simeq$ where $\mathbf{T}(2,4)$ is a certain "track category". Moreover $\mathbf{T}(2,4)$ can be described as the pull back of two quotient categories $\mathbf{TL} = \mathbf{T}(2,4)/E_\Gamma$ and $\mathbf{T\Gamma} = \mathbf{T}(2,4)/E_L$. In chapters 4 and 5 we shall describe explicit algebraic models of these categories $\mathbf{TL}$ and $\mathbf{T\Gamma}$ respectively. This leads to an algebraic model of the category $\mathbf{CW}(2,4)$ in chapter 5.

3.1. The track category of one point unions of n-spheres

Let I be the unit intervall and let I_*X be the (reduced) cylinder of X, that is $I_*X = (I \times X)/(I \times *)$ where $* \in X$ is the basepoint. Let $f, g : X \to Y$ be maps in $\mathbf{Top}^*$. A *track* $H : f \simeq g$ is a homotopy class relative to $X \vee X \subset I_*X$ of homotopies $I_*X \to Y$ from f to g. Let

$$(3.1.1) \qquad T(f,g) = [I_*X, Y]^{(f,g)}$$

be the set of such tracks. This set is defined similarly in any cofibration category, see [BAH]. For $S^1 = I/\partial I$ and $\Sigma_*X = (S^1 \times X)/(S^1 \times *)$ we have a homotopy equivalence under X

$$(1) \qquad \Sigma_*X \simeq X \vee \Sigma X$$

in case $X = \Sigma X$ is a suspension.This shows that one has an isomorphism of groups

(2) $$\sigma_f : T(f,f) = [\Sigma_* X, Y]^f \cong [\Sigma X, Y].$$

We use σ_f for the transitive and effective action

(3) $$T(f,g) \times [\Sigma X, Y] \xrightarrow{+} T(f,g)$$

defined by $H + \alpha = H + \sigma_g^{-1}(\alpha)$. Here $H + \sigma_g^{-1}(\alpha) = \sigma_f^{-1}(\alpha) + H$ is given by track addition, see [BAH] or [BHC]. Compare also the papers on "linear track extensions" [BD], [BJA], [BJR]. We now consider tracks as in (3.1.1) where X and Y are Moore spaces of free abelian groups, such tracks arise for example in the proof of (2.5). We first introduce some notation.

3.1.2. Definition: For a free abelian group A with a (fixed) basis Z let

(1) $$M_A = \bigvee_Z S^1, \; M(A,n) = \Sigma^{n-1} M_A,$$

(2) $$G_A = \pi_1(M_A) =< Z >,$$

(3) $$E_A = G_A / \Gamma_3 G_A.$$

Here G_A is the *free group* generated by Z and E_A is the *free nil*(2)*-group* generated by Z; the group $\Gamma_3 G_A$ is the subgroup generated by all triple commutators in G_A. A map $x : M_{A'} \to M_A$ induces via $\pi_1(x)$ a homomorphism

(4) $$\xi = x_* : E_{A'} \to E_A$$

For a group G let $G^{ab} = G/\Gamma_2 G$ be the *abelianization* of G; here $\Gamma_2 G$ denotes the commutator subgroup of G. An element $x \in G$ represents $\{x\} \in G^{ab}$. The group structure of a group $(G,+)$ is written *additively* though the group law $+$ need not be abelian. Thus $-x-y+x+y = (x,y)$ is the *commutator* in G and $0 \in G$ is the neutral element. We clearly have the abelianization $G_A^{ab} = E_A^{ab} = A$, see (4.2), and there is the well known central extension

(3.1.3) $$A \wedge A \overset{w}{\hookrightarrow} E_A \overset{p}{\twoheadrightarrow} A$$

with $p(x) = \{x\}$ and $w(\{x\} \wedge \{y\}) = (x, y)$ for $x, y \in E_A$. For homomorphisms $\xi, \eta : E_A \to E_A$ with $\xi^{ab} = \eta^{ab}$ there is a unique $\Delta = \Delta(\xi, \eta) \in Hom(A, A \wedge A)$ with $\xi + w\Delta p = \eta$. We define the subsets

$$T_2(\xi, \eta) = \{\alpha \in Hom(A, A \otimes A) : \; q\alpha = \Delta(\xi, \eta)\}, \tag{1}$$

$$T_n(\xi, \eta) = \{\alpha \in Hom(A, A \hat{\otimes} A) : \; q\alpha = \Delta(\xi, \eta)\}, \tag{2}$$

where $n \geq 3$. Here q is the projection in (1.1.2). The elements in $T_n(\xi, \eta)$ are algebraic models of certain tracks since we have the following result.

3.1.4. PROPOSITION: *Let $x, y : M_A \to M_A$ be maps which induce $\xi, \eta : E_{A'} \to E_A$, that is $\xi = x_*$, $\eta = y_*$ as in (3.1.2)(4). Then there is a natural bijection ($n \geq 2$)*

$$\chi : \; T(\Sigma^{n-1}x, \Sigma^{n-1}y) \equiv T_n(\xi, \eta).$$

We prove this result for $n = 2$ in (6.4.2), compare also VI.4.7 in [BCH] and §5 of [BD] The bijection χ carries addition of tracks to the corresponding addition of homomorphisms. Moreover χ is compatible with the action in (3.1.1) in the sense that

$$\chi(H + \alpha) \; = \; \chi(H) + \tau\alpha \tag{1}$$

where τ is the inclusion in (1.1.2) and where we use the identification

$$[M(A', n+1), M(A, n)] \; = \; Hom(A', \Gamma^1_n A) \tag{2}$$

with $\Gamma^1_2(A) = \Gamma(A)$ and $\Gamma^1_n(A) = A \otimes \mathbb{Z}/2$ for $n \geq 3$, compare chapter 1. We also use χ in (3.1.4) as an identification. In case $\xi = \eta$ one gets *the canonical track*

$$\mathcal{O} : \Sigma^{n-1}x \simeq \Sigma^{n-1}y \tag{3}$$

which corresponds to the trivial homomorphism $\chi(\mathcal{O}) = 0$ in $T_n(\xi, \xi)$.

3.2. The linear extension $\mathbf{T}(2,4)$ defined by tracks

Using tracks we define a linear extension

$$E \to \mathbf{T}(2,4) \to \mathbf{E}(2,4)$$

where $\mathbf{E}(2,4)$ is an algebraic category and E is an algebraically defined natural system on $\mathbf{E}(2,4)$. The category $\mathbf{T}(2,4)$, however, is defined in terms of spaces and continuous maps. It is the main purpose of the following chapters to compute $\mathbf{T}(2,4)$ algebraically. There is a natural equivalence relation $\simeq$ on $\mathbf{T}(2,4)$ auch that the homotopy category $\mathbf{CW}(2,4)$ of (2,4)-complexes is equivalent to the category

$$\mathbf{CW}(2,4) = \mathbf{T}(2,4)/\simeq$$

This way we shall otain an algebraic model category of $\mathbf{CW}(2,4)$. Recall that by (1.3.8) we have the linear extension

$$D \to \mathbf{CW}(2,4) \to \mathbf{H}(2,4).$$

We now define the category $\mathbf{E}(2,4)$ together with an abelianization functor

$$ab : \mathbf{E}(2,4) \to \mathbf{H}(2,4) \tag{3.2.1}$$

Objects in $\mathbf{E}(2,4)$ are the same as in $\mathbf{H}(2,4)$, namely homomorphisms $g : B \to \Gamma(A)$ where A and B are free abelian groups. Morphisms $(\xi, \eta) : f \to g$ in $\mathbf{E}(2,4)$ are given by homomorphisms $\xi : E_{B'} \to E_B$ and $\eta : E_{A'} \to E_A$, see (3.1.2), for which the diagram

$$\begin{array}{ccc} B' & \xrightarrow{\xi^{ab}} & B \\ {\scriptstyle f}\downarrow & & \downarrow{\scriptstyle g} \\ \Gamma(A') & \xrightarrow{\Gamma(\eta^{ab})} & \Gamma(A) \end{array} \tag{1}$$

commutes. The functor (3.2.1) carries (ξ, η) to (ξ^{ab}, η^{ab}). Composition in $\mathbf{E}(2,4)$ is defined by $(\xi, \eta)(\xi', \eta') = (\xi\xi', \eta\eta')$.

We associate with $\mathbf{E}(2,4)$ a topological *track category* $\mathbf{T}(2,4)$ as follows. For this we choose for each object g in $\mathbf{E}(2,4)$ a map

$$g : M(B,3) \to M(A,2) \tag{2}$$

in **Top*** as in (1.2.1). Moreover we choose for each homomorphism $\rho : E_{A'} \to E_A$ a map $t(\rho) : M_{A'} \to M_A$ which induces $\rho = t(\rho)_*$, see (3.1.2)(4). The objects of $\mathbf{T}(2,4)$ are the maps in (2) which we identify with the objects in $\mathbf{E}(2,4)$. The morphisms $f \to g$ in $\mathbf{T}(2,4)$ are triples (ξ, η, H) where $(\xi, \eta) : f \to g$ is a morphism in $\mathbf{E}(2,4)$ and where H is a track as in the diagram

(3)
$$\begin{array}{ccc} M(B',3) & \xrightarrow{\Sigma^2 t\xi} & M(B,3) \\ {\scriptstyle f}\downarrow & \overset{H}{\Rightarrow} & \downarrow{\scriptstyle g} \\ M(A',2) & \xrightarrow[\Sigma t\eta]{} & M(A,2) \end{array}$$

Such a track exists since (1) commutes. Composition of such morphisms is defined by the tracks in the following diagram:

(4)
$$\begin{array}{ccccc} & & \overset{\Sigma^2 t(\xi\xi')}{\frown} & & \\ M(B'',3) & \xrightarrow[\Sigma^2 t\xi']{} & M(B',3) & \xrightarrow[\Sigma^2 t\xi]{} & M(B,3) \\ \downarrow & & \downarrow & & \downarrow \\ M(A'',2) & \xrightarrow{\Sigma t\eta'} & M(A',2) & \xrightarrow{\Sigma t\eta} & M(A,2) \\ & & \underset{\Sigma t(\eta\eta')}{\smile} & & \end{array}$$

Here we use the canonical track $\mathcal{O}$ in (3.1.4)(3). Naturality of χ in (3.1.4) shows that $\mathbf{T}(2,4)$ is a well defined category. Moreover using the action (3.1.1) we obtain a linear extension of categories

(5)
$$E+ \rightarrowtail \mathbf{T}(2,4) \xrightarrow{p} \mathbf{E}(2,4)$$

where p is the forgetful functor and where the natural system E on $\mathbf{E}(2,4)$ is a bimodule given by

(6)
$$E(\xi,\eta) = Hom(B', \Gamma_2^2(A)) = [M(B',4), M(A,2)].$$

Using (3.1.1) we see that $\alpha \in E(\xi, \eta)$ acts on morphisms $f \to g$ in $\mathbf{T}(2,4)$ by $(\xi, \eta, H) + \alpha = (\xi, \eta, H + \alpha)$. It is not hard to see that this action yields a linear extension of categories as in (5). The crucial property of the track category $\mathbf{T}(2,4)$ is described by the following commutative diagram in which the rows are linear extensions:

(3.2.2)
$$\begin{array}{ccccc} E+ & \rightarrowtail & \mathbf{T}(2,4) & \stackrel{p}{\twoheadrightarrow} & \mathbf{E}(2,4) \\ \downarrow q & & \downarrow Q & & \downarrow ab \\ D+ & \rightarrowtail & \mathbf{CW}(2,4) & \stackrel{H_*}{\twoheadrightarrow} & \mathbf{H}(2,4) \end{array}$$

Here the functor Q carries the object g in (3.2.1)(2) to the mapping cone of g and carries the morphism (ξ, η, H) in (3.2.1)(3) to the principal map between mapping cones associated with the track H. The functor Q is full, compare the proof of (1.3.8). Moreover q is the quotient map of natural systems given by (1.3.8)(5). One can check that Q is equivariant in the sense that $Q(\xi, \eta, H + \alpha) = Q(\xi, \eta, H) + q(\alpha)$. We now define a natural equivalence relation $\simeq$ on the category $\mathbf{T}(2,4)$ as follows. We set $(\xi, \eta, H) \simeq (\xi', \eta', H')$ if there exist homotopies

(1)
$$H_\eta \in T_2(\eta, \eta'),\ \ H_\xi \in T_3(\xi, \xi')$$

(see (3.1.4)), such that $\alpha : B' \to \Gamma_2^2(A)$ with

(2)
$$-f^* H_\eta + H + g_* H_\xi = H' + \alpha$$

satisfies $(i_g)_* \alpha = 0$. Here $(i_g)_* : \Gamma_2^2(A) \to \Gamma_2^2(g)$ is the map in (1.2.5).

3.2.3. PROPOSITION: *The functor Q in (3.2.2) induces an equivalence of categories*

$$\mathbf{T}(2,4)/\simeq\ \tilde{\to} \mathbf{CW}(2,4).$$

We leave the proof as an exercise; compare II.13.9 of [BAH].

In the next two chapters we compute an algebraic category which is equivalent to the category $\mathbf{T}(2,4)$. Moreover, we compute the equivalence relation $\simeq$ on this algebraic category. This way we obtain an algebraic model category of $\mathbf{CW}(2,4)$.

3.2.4. REMARK: Let $\mathbf{CW}(n,m)$, $2 \le n \le m$, be the full homotopy category in $\mathbf{Top}^*/\simeq$ consisting of CW-complexes X with cells only in dimensions n and m. We can define the track category $\mathbf{T}(n,m)$ similarly to (3.2.1)(3) and one has for $\mathbf{T}(n,m) \to \mathbf{CW}(n,m)$ a diagram of linear extensions of categories as in (3.2.2). Moreover the analogue of (3.2.3) holds for $\mathbf{CW}(n,m)$. The computation of the categories $\mathbf{T}(n,m)$ and $\mathbf{CW}(n,m)$ remains a challenging problem. In this book we only deal with the case $(n,m)=(2,4)$. Also the case $(n,m)=(n,n+1)$ is known by (3.1.4). For the computation of $\mathbf{T}(n,m)$ compare [BNA].

3.3. The linear extensions $\mathbf{T\Gamma}$ and $\mathbf{TL}$

The splitting $\Gamma_2^2(A)=\Gamma(A)\otimes\mathbb{Z}/2\oplus L(A,1)_3$ yields canonically linear extensions $\mathbf{T\Gamma}$ and $\mathbf{TL}$ such that $\mathbf{T}(2,4)$ in (3.2) is a pull back of

$$\mathbf{T\Gamma}\to\mathbf{E}(2,4)\leftarrow\mathbf{TL}.$$

We have seen in (1.1.9) that the functor Γ_2^2 splits as a direct sum of two functors. This implies that the natural system E on $\mathbf{E}(2,4)$ in (3.2.1)(5) splits as well, that is $E=E_\Gamma\oplus E_L$ with

$$\text{(3.3.1)}\qquad \begin{cases} E_\Gamma(\xi,\eta) &= Hom(B',\Gamma(A)\otimes\mathbb{Z}/2),\\ E_L(\xi,\eta) &= Hom(B',L(A,1)_3).\end{cases}$$

The cohomology class of the extension (3.2.1)(5) thus splits as a sum of two elements since for $\mathbf{E}=\mathbf{E}(2,4)$ we have the isomorphism

$$\text{(1)}\qquad \begin{cases} H^2(\mathbf{E},E) &\cong H^2(\mathbf{E},E_\Gamma)\oplus H^2(\mathbf{E},E_L)\\ <\mathbf{T}(2,4)> &\mapsto <\mathbf{T\Gamma}>\oplus<\mathbf{TL}>\end{cases}$$

We obtain the linear extensions

$$\text{(2)}\qquad E_\Gamma+\hookrightarrow\mathbf{T\Gamma}\twoheadrightarrow\mathbf{E},$$

$$\text{(3)}\qquad E_L+\hookrightarrow\mathbf{TL}\twoheadrightarrow\mathbf{E}$$

simply by the quotient categories $\mathbf{T\Gamma}=\mathbf{T}(2,4)/E_L$ and $\mathbf{TL}=\mathbf{T}(2,4)/E_\Gamma$. These quotients are obtained by dividing out the actions of $E_L\subset E$ and $E_\Gamma\subset E$ as follows.

3.3.2. DEFINITION: Let $D\to\mathbf{E}\xrightarrow{p}\mathbf{C}$ be a linear extension and let $F\subset D$ be an inclusion of natural systems on $\mathbf{C}$. Then we obtain the linear extension

$$D/F\to\mathbf{E}/F\to\mathbf{C}$$

as follows. For $f,g:A\to B$ in $\mathbf{E}$ we write $f\sim g$ if $pf=pg$ in $\mathbf{C}$ and if there is $\alpha \in F(pf)$ with $g = f + \alpha$. Then $\sim$ is a natural equivalence relation on $\mathbf{E}$ which defines the quotient category $\mathbf{E}/\sim = \mathbf{E}/F$.

Using the linear extension (3.3.1)(2),(3) we obtain the *pull back* of categories

(3.3.3)
$$\begin{array}{ccc} \mathbf{T}(2,4) & \xrightarrow{q} & \mathbf{T}\Gamma \\ {\scriptstyle q}\downarrow & & \downarrow \\ \mathbf{TL} & \xrightarrow[p]{} & \mathbf{E}(2,4) \end{array}$$

Here q denotes the quotient functor. One of the main results below shows that the extension $\mathbf{TL}$ is split. The computation of $\mathbf{T}\Gamma$ is fairly complicated.

CHAPTER 4

THE SPLITTING OF THE LINEAR EXTENSION TL

We introduce an algebraic category **LT** which is equivalent to the topologically defined category **TL** in (3.3.1)(3). This result is proved in chapter 6. We here study the algebraic category **LT** in terms of extension groups of functors. The computation of such extension groups yields the result that **LT**, and hence **TL**, is a split linear extension of categories.

4.1. The quadratic refinement $\tilde{\Gamma}$ and an algebraic model of the track category TL

We introduce in this section a functor $\tilde{\Gamma}$ which is a refinement of Whitehead's quadratic functor Γ. Recall that Γ is defined by a universal property with respect to the distributivity laws of a quadratic function. Similarly we define the functor $\tilde{\Gamma}$ by a universal construction with respect to more sophisticated distributivity laws. Using the refinement $\tilde{\Gamma}$ we describe an algebraic category **LT** which is equivalent to the topological category $\mathbf{TL} = \mathbf{T}(2,4)/E_\Gamma$. The results in this section are proved in chapter 6 by use of crossed chain complexes.

Recall the group law of a group G is written additively though G need not be an abelian group. We now define the functor $\tilde{\Gamma}$ as follows.

4.1.1. DEFINITION: Let G be a group (not necessarily abelian) and let A be an abelian group. We call a function f and a homomorphism F as in the diagram

$$G \xrightarrow{f} A \xleftarrow{F} \otimes^3 G^{ab}$$

a *G-quadratic-pair* if the following properties are satisfied. For $x, y, z \in G$ we write

$$[x, y, z] = F(\{x\} \otimes \{y\} \otimes \{z\})$$

where $\{x\} \in G^{ab}$ is represented by x. Then f and F satisfy the following equations (1)...(4) where we set

$$[x, y] = f(x + y) - f(x) - f(y):$$

(1) $$[x, x, y] = [y, x, x],$$

(2) $$f(-x) = f(x) + 2[x, x, x],$$

(3) $$[x + y, z] = [x, z] + [y, z] - \Delta(x, y, z),$$

(4) $$[x, y + z] = [x, y] + [x, z] - \Delta(x, y, z),$$

where $\Delta(x, y, z) = [x, z, y] + 2[z, y, x] + 3[z, x, y]$. A G-quadratic pair

(5) $$G \xrightarrow{\gamma} \tilde{\Gamma} \xleftarrow{\delta} \otimes^3 G^{ab}$$

is *universal* if for each G-quadratic pair (f, F) as above there is a unique homomorphism

(6) $$(f, F)^{\square} : \tilde{\Gamma}(G) \to A$$

with $(f, F)^{\square}\gamma = f$ and $(f, F)^{\square}\delta = F$. The universal quadratic pair exists and yields a functor

(7) $$\tilde{\Gamma} : \mathbf{Gr} \to \mathbf{Ab}.$$

For a homomorphism $\alpha : G' \to G$ the induced homomorphism $\tilde{\Gamma}(\alpha)$ is given by $\tilde{\Gamma}(\alpha) = (\gamma\alpha, \delta \otimes^3 \alpha^{ab})^{\square}$.

We call the functor $\tilde{\Gamma}$ a *refinement* of Whitehead's functor Γ since the equations (2), (3), (4) correspond to the equations of a quadratic map in (1.1.1). In fact, a G-quadratic pair (f, F) is the same as a quadratic map f in case $F = 0$ is trivial. We now describe some basic properties of the functor $\tilde{\Gamma}$ which are proved in chapter 6. We first observe that $\tilde{\Gamma}(G)$ actually depends only on the quotient $G/\Gamma_3 G$ where $\Gamma_3 G \subset G$ is the subgroup of all triple commutators in G with the quotient map $q : G \to G/\Gamma_3 G$.

4.1.2. PROPOSITION: *The induced map*

$$q_* : \tilde{\Gamma}(G) \to \tilde{\Gamma}(G/\Gamma_3 G)$$

is an isomorphism.

This is a consequence of the next lemma which is proved in (6.1.3).

4.1.3. LEMMA: *Let (f, F) be a G-quadratic pair. Then we have for $x, y, z \in G$*

$$\begin{aligned} f(-x-y+x+y) &= 0, \\ f(z-x-y+x+y) &= f(z) + [x,y,z] - [z,x,y]. \end{aligned}$$

Our main result on the functor $\tilde{\Gamma}$ is the following theorem.

4.1.4. THEOREM: *For any group G there is a natural sequence of homomorphisms*

$$0 \to L(G^{ab}, 1)_3 \xrightarrow{i} \otimes^3 G^{ab} \xrightarrow{\delta} \tilde{\Gamma}(G) \xrightarrow{\tilde{p}} \Gamma(G^{ab}) \to 0$$

which satisfy $\delta i = 0$ and $\tilde{p}\delta = 0$. Moreover if G is a free group, or if G is a free nil(2)-group, then the sequence is exact.

The natural map i is the inclusion of triple Lie brackets in (1.1.8)(2) and $\delta i = 0$ follows from (4.1.1)(1), see (6.1.4). Moreover $\tilde{p}$ is obtained by the G-quadratic pair $(\gamma h, 0)$ where $\gamma h : G \to G^{ab} \to \Gamma(G^{ab})$ is given by the quotient map h and by γ in (1.1.1). We set $\tilde{p} = (\gamma h, 0)^{\square}$; this implies $\tilde{p}\delta = 0$. We shall prove the complicated part of (4.1.4), namely the exactness of the sequence for a free group G in (6.3.1) and (6.2.9).

Recall that for a free abelian group A the free nil(2)-group E_A is the quotient $E_A = G/\Gamma_3 G$ where G is the free group with $G^{ab} = E_A^{ab} = A$, see (3.1.2). By theorem (4.1.4) we obtain an exact sequence

(4.1.5)
$$L(A,1)_3 \rightarrowtail \otimes^3 A \xrightarrow{\delta} \tilde{\Gamma}(E_A) \xrightarrow{\tilde{p}} \Gamma(A)$$

which is natural for homomorphisms $\eta : E_{A'} \to E_A$ between free nil(2)-groups. The induced homomorphisms $\tilde{\Gamma}(\eta)$ have the following property. Let $\eta' : E_{A'} \to E_A$ be given with $(\eta')^{ab} = \eta^{ab}$, then we can choose

(1)
$$H_\eta \in T_2(\eta, \eta'),\ H_\eta : A' \to A \otimes A,$$

as in (3.1.3)(1). Consider the diagram

(2)
$$\begin{array}{ccc} \tilde{\Gamma}(E_{A'}) & \xrightarrow{\tilde{\Gamma}(\eta),\tilde{\Gamma}(\eta')} & \tilde{\Gamma}(E_A) \\ {\scriptstyle \tilde{p}}\downarrow & & \uparrow{\scriptstyle \delta} \\ \Gamma(A') & \xrightarrow{\tilde{H}} & \otimes^3 A \end{array}$$

Here $\tilde{H}_\eta$ is defined by the sum

(3)
$$\tilde{H}_\eta = (\eta^{ab} \otimes H_\eta - H_\eta \otimes \eta^{ab})\tau$$

where we use $\tau : \Gamma(A') \to A' \otimes A'$. Now lemma (4.1.3) implies the formula

(4)
$$-\tilde{\Gamma}(\eta) + \tilde{\Gamma}(\eta') = \delta \tilde{H}_\eta \tilde{p}.$$

The exact sequence in (4.1.5) is the crucial ingredient in the next definition of an "algebraic track category". Recall that $\mathbf{E}(2,4)$ is defined in (3.2.1).

4.1.6. DEFINITION: Let **LT** be the following algebraic category which is part of a linear extension:

(1)
$$E_L + \rightarrowtail \mathbf{LT} \xrightarrow{p} \mathbf{E}(2,4).$$

For each object $g : B \to \Gamma(A)$ in $\mathbf{E}(2,4)$ we fix a homomorphism $\tilde{g} : B \to \tilde{\Gamma}(E_A)$ with $\tilde{p}\tilde{g} = g$. We do this by fixing for each free abelian group A a homomorphism $\tilde{g}_A : \Gamma(A) \to \tilde{\Gamma}(E_A)$ with $\tilde{p}\tilde{g}_A = 1_{\Gamma A}$. Then we set $\tilde{g} = \tilde{g}_A g$:

$B \to \Gamma(A) \to \tilde{\Gamma}(E_A)$. We consider $\tilde{g}$ as an object in **LT** which is identified by the functor p in (1) with the object g. A morphism $\tilde{f} \to \tilde{g}$ in **LT** is a triple (ξ, η, H) where $(\xi, \eta) : f \to g$ is a morphism in $\mathbf{E}(2,4)$, see (3.2.1), and where H is a homomorphism

(2)
$$\begin{cases} H : B' \to \otimes^3 A \text{ with} \\ \delta H = -\tilde{\Gamma}(\eta)\tilde{f} + \tilde{g}(\xi^{ab}). \end{cases}$$

For this we consider the diagram

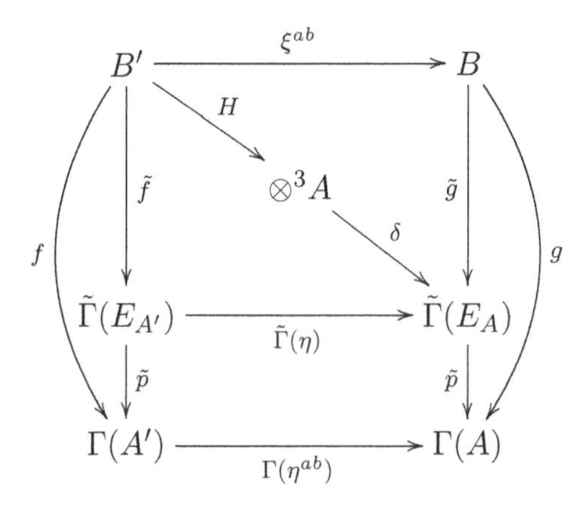

Since $g(\xi^{ab}) = \Gamma(\eta^{ab})f$ by (4.5), we see that $\tilde{p}(-\tilde{\Gamma}(\eta)\tilde{f} + \tilde{g}(\xi^{ab})) = 0$. Hence by exactness of (4.1.5) there exists H satisfying (2). This shows that the forgetful functor p in (1) which carries (ξ, η, H) to (ξ, η) is full. We define the composition in **LT** by

(3)
$$\begin{cases} (\xi, \eta, H)(\xi', \eta', H') = (\xi\xi', \eta\eta', H * H') \text{ with} \\ H * H' = H((\xi')^{ab}) + (\otimes^3 \eta^{ab})H'. \end{cases}$$

Finally we obtain the action of $\alpha \in E_L(\xi, \eta) = Hom(B', L(A,1)_3)$ by

(4)
$$(\xi, \eta, H) + \alpha = (\xi, \eta, H + i\alpha)$$

where i is the inclusion in (4.1.5). Using the exactness on (4.1.5) one readily checks that (4) yields a well defined linear extension of categories as described in (1).

The next result shows that the category **LT** is an algebraic model of the topological track category **TL** in (3.3.1)(3).

4.1.7. THEOREM: *There is an isomorphism*

$$\chi : \mathbf{TL} \overset{\cong}{\to} \mathbf{LT}$$

of categories which yields an equivalence of the linear extensions (4.1.6)(1) and (3.3.1)(3).

This result is proved in (6.4.4) below. In addition we get the following result which approximates the homotopy category $\mathbf{CW}(2,4)$. Recall that we have a surjection of natural systems $E \twoheadrightarrow D$ given by the map

$$q : E(\xi,\eta) = Hom(B', \Gamma_2^2 A) \twoheadrightarrow D(\xi,\eta) \tag{4.1.8}$$

see (1.3.3). Let $D_\Gamma = qE_\Gamma$ and $D_L = qE_L$ be the corresponding images of E_Γ and E_L respectively, see (3.3.1). Then D_Γ and D_L are natural subsystems of D. Clearly D_Γ consits only of $\mathbb{Z}/2$-vector spaces. The next result is a consequence of (3.2.3) and (6.4.6) below.

4.1.9. THEOREM: *There is an equivalence of linear extensions*

$$\mathbf{CW}(2,4)/D_\Gamma \overset{\sim}{\to} \mathbf{LT}/\simeq$$

where the notion of homotopy $\simeq$ on **LT** *is defined in (4.1.10) below..*

We describe a further model category of $\mathbf{CW}(2,4)/D_\Gamma$ in (4.5) below.

4.1.10. DEFINITION: We set $(\xi,\eta,H) \simeq (\xi',\eta',H')$ if $\xi^{ab} = (\xi')^{ab}$, $\eta^{ab} = (\eta')^{ab}$ and if there is a homotopy

$$H_\eta \in T_2(\eta,\eta'),\ H_\eta : A' \to A \otimes A,$$

as in (4.4) such that the homomorphism $\alpha : B' \to L(A,1)_3$ which we define below satisfies $q(\alpha) = 0$ in $D(\xi,\eta)/D_\Gamma(\xi,\eta)$, see (4.1.8). We obtain α similarly as in (3.2.2)(2) by the formula

$$-\tilde{H}_\eta f + H = H' + i\alpha. \tag{1}$$

Here i is the inclusion in (4.1.5) and $\tilde{H}_\eta$ is defined in (4.1.5)(3). Formula

(4.1.5)(4) implies by the assumption on H, H', H_η that

$$(2) \qquad \delta(-\tilde{H}_\eta f + H) = \delta H'$$

so that α in (1) is well defined.

4.2. Extension of functors

Let $\mathbf{C}$ be a (small) category. We consider the functor category $Fun(\mathbf{C}, \mathbf{Ab})$ of functors from $\mathbf{C}$ to the category of abelian groups $\mathbf{Ab}$. Morphisms are natural transformations. Since $Fun(\mathbf{C}, \mathbf{Ab})$ is an abelian category we can define the extension groups $Ext^n_{\mathbf{C}}(F, G)$ in $Fun(\mathbf{C}, \mathbf{Ab})$ in the ususal way. An element in $Ext^n_{\mathbf{C}}(F, G)$ is represented by an exact sequence in $\mathbf{Ab}$

$$\mathcal{E} : 0 \to G(X) \to F_1(X) \to \cdots \to F_n(X) \to F(X) \to 0$$

which is natural in $X \in \mathbf{C}$. Two such sequences are Yoneda equivalent if there is a commutative diagram

$$\begin{array}{cccccccccccc}
\mathcal{E} & 0 \longrightarrow & G(X) & \longrightarrow & F_1(X) & \longrightarrow \cdots \longrightarrow & F_n(X) & \longrightarrow & F(X) & \longrightarrow 0 \\
 & & \| & & \downarrow & & \downarrow & & \| & \\
\mathcal{E}' & 0 \longrightarrow & G(X) & \longrightarrow & G_1(X) & \longrightarrow \cdots \longrightarrow & G_n(X) & \longrightarrow & F(X) & \longrightarrow 0
\end{array}$$

Yoneda equivalence generates an equivalence relation for exact sequences and $Ext^n_{\mathbf{C}}(F, G)$ is the set of such equivalence classes.

For example let $\mathbf{C} = \mathbf{nil}$ be a full category of free nil(2)-groups. We assume that $\mathbf{C}$ is a small category so that the objects of $\mathbf{nil}$ form a set, for example we can define $\mathbf{nil}$ by the objects E_A where $A = \mathbb{Z}^n$, $n \geq 0$, is a finitely generated abelian group as in (3.1.2). Using the abelianization functor $\mathbf{nil} \to \mathbf{ab}$ where $\mathbf{ab}$ is the category of free abelian groups we get the composite functors

$$(4.2.1) \qquad \Gamma,\ L(\_, 1)_3 : \mathbf{nil} \to \mathbf{ab} \to \mathbf{Ab}$$

which carry E_A to $\Gamma(A)$ and $L(A, 1)_3$ respectively. Hence the natural exact sequence (4.1.5) represents an element in $Ext^2_{\mathbf{nil}}(\Gamma, L(\_, 1)_3)$. We shall prove in (4.3) below

4.2.2. THEOREM: *The element* $<\mathcal{L}>\in Ext^2_{\mathbf{nil}}(\Gamma, L(\_,1)_3)$ *represented by the natural exact sequence*

$$\mathcal{L}: 0 \to L(A,1)_3 \to \otimes^3 A \to \tilde{\Gamma}(E_A) \to \Gamma(A) \to 0$$

in (4.1.5) is trivial, that is $<\mathcal{L}>=0$.

This result actually implies that the linear extension **LT** in (4.1.6) is split. Hence by (4.1.7) also **TL** is split. To see this we use the following connection between elements in $Ext^2_{\mathbf{C}}(F,G)$ and linear extensions.

4.2.3. DEFINITION: Let

$$\mathcal{E}: 0 \to G \xrightarrow{i} F_2 \xrightarrow{\delta} F_1 \xrightarrow{\tilde{p}} F \to 0$$

be an exact sequence in $Fun(\mathbf{C}, \mathbf{Ab})$ and assume $F(X)$ is free abelian for all objects X in $\mathbf{C}$. Then we associate with $\mathcal{E}$ a linear extension

$$Hom(F,G) \to \mathbf{E}(\mathcal{E}) \to \mathbf{C}$$

as follows. Objects in $\mathbf{E}(\mathcal{E})$ are objects X in $\mathbf{C}$ for which we choose a homomorphism

$$\tilde{g}_X : F(X) \to F_1(X)$$

which is a splitting of $\tilde{p}: F_1(X) \to F(X)$. This is possible since we assume $F(X)$ to be free abelian. Morphisms $Y \to X$ in $\mathbf{E}(\mathcal{E})$ are pairs (η, H) where $\eta: Y \to X$ is a morphism in $\mathbf{C}$ and $H: F(Y) \to F_2(X)$ is a homomorphism satisfying

$$\delta(H) = -F_1(\eta)\tilde{g}_Y + \tilde{g}_X F(\eta).$$

The natural system $Hom(F,G)$ on $\mathbf{C}$ carries $\eta: Y \to X$ to $Hom(F(Y), G(X))$. An element $\alpha \in Hom(F(Y), G(X))$ acts on $(\eta, H): Y \to X$ by $(\eta, H) + \alpha = (\eta, H + i\alpha)$.

The next result is readily checked.

4.2.4. PROPOSITION: *For* X *in* $\mathbf{C}$ *let* $F(X)$ *be free abelian. Then the linear extension* $\mathbf{E}(\mathcal{E})$ *is well defined and a Yoneda equivalence* $\mathcal{E} \to \mathcal{E}'$ *yields an equivalence of linear extensions* $\mathbf{E}(\mathcal{E}) \to \mathbf{E}(\mathcal{E}')$. *Hence the function*

$$\lambda : Ext^2_{\mathbf{C}}(F, G) \to H^2(\mathbf{C}, Hom(F, G))$$

which carries $< \mathcal{E} >$ *to* $\{\mathbf{E}(\mathcal{E})\}$ *is well defined; see (1.3.9).*

Jibladze and Pirashvili [JP] show that λ in (4.2.4) is an isomorphism of abelian groups. Such an isomorphism exists for all degrees Ext^n, H^n with $n \in \mathbb{Z}$.

Next we show that the linear extension $\mathbf{LT}$ in (4.1.6) can be expressed in terms of the linear extension $\mathbf{E}(\mathcal{L})$ defined in (4.2.3) by the extension $\mathcal{L}$ in (4.2.2).

4.2.5. PROPOSITION: *There is a pullback diagram of linear extensions with* $L = L(\_, 1)_3$

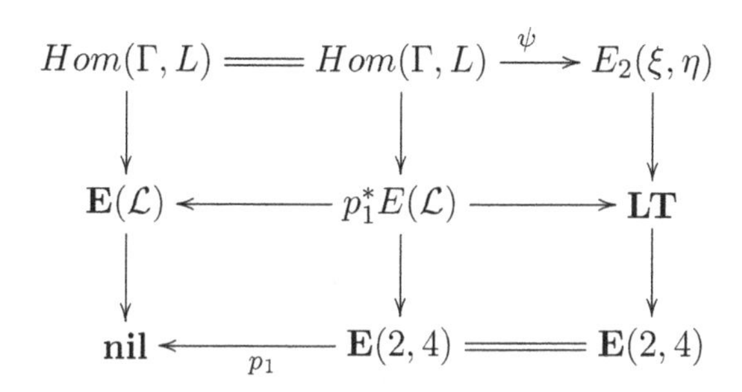

Here p_1 is the forgetful functor which carries the object $g : B \to \Gamma(A)$ in $\mathbf{E}(2,4)$ to the group E_A and carries $(\xi, \eta) : f \to g$ with $f : B' \to \Gamma(A')$ in $\mathbf{E}(2,4)$ to η; compare (3.2.1). Moreover the map ψ between natural systems is given by the commutative diagram (see (3.3.1))

$$\begin{array}{ccc} Hom(\Gamma, L)(\xi, \eta) & \xrightarrow{\psi} & E_L(\xi, \eta) \\ \| & & \| \\ Hom(\Gamma A', LA) & \xrightarrow[f^*]{} & Hom(B', LA) \end{array}$$

Theorem (4.2.2) and (4.2.5) yield the following result on the splitting of $\mathbf{LT}$ and $\mathbf{TL}$.

4.2.6. THEOREM: *The linear extensions* $\mathbf{LT}$ *and* $\mathbf{TL}$ *are split.*

Proof: We have the cohomology classes

$$\{\mathbf{TL}\} = \{\mathbf{LT}\} = \psi_* p_1^* \{\mathbf{E}(\mathcal{L})\} = \psi_* p_1^* \lambda(< \mathcal{L} >)$$

where $< \mathcal{L} >= 0$ by (4.4.2). The second equation holds by (4.2.5). □

4.3. The extension class $< \mathcal{L} >$

For a free abelian group A we have the composite

$$[[1,1],1]i : L(A,1)_3 \to \otimes^3 A \twoheadrightarrow L(A,1)_3.$$

Compare (1.1.8)(2).

4.3.1. LEMMA: *The composite* $[[1,1],1]i$ *is multiplication by* 3; *that is for* $v \in L(A,1)_3$ *we have* $[[1,1],1]i(v) = 3 \cdot v \in L(A,1)_3$.

Proof: $L(A,1)_3$ is generated by elements of the form $[[x,y],z]$. For such elements we have

$$\begin{aligned} [[1,1],1]i[[x,y],z] &= [[1,1],1]((xy+yx)z - z(xy+yx)) \\ &= [[x,y],z] + [[y,x],z] - [[z,x],y] - [[z,y],x] \\ &= 3[[x,y],z]. \end{aligned}$$

Here we use $[x,y] = [y,x]$ and the Jacobi identity

$$[[x,y],z] + [[z,x],y] + [[y,z],x] = 0. \quad \square$$

The extension $\mathcal{L}$ in (4.1.5) leads to the following commutative diagram where push denotes a push out diagram of abelian groups. Let A be a free abelian group and $L(A) = L(A,1)_3$.

$$\begin{array}{ccccccccc} LA & \rightarrowtail & \otimes^3 A & \twoheadrightarrow & ker(\tilde{p}) & \rightarrowtail & \tilde{\Gamma}(E_A) & \rightarrow & \Gamma A \\ \| & & \downarrow{\scriptstyle [[1,1],1]} & & \downarrow & & \downarrow{\scriptstyle q} & & \| \\ LA & \overset{\cdot 3}{\rightarrowtail} & LA & \twoheadrightarrow & \mathbb{Z}/3 \otimes LA & \rightarrowtail & \Gamma_{\mathbb{Z}/3}(E_A) & \twoheadrightarrow & \Gamma A \end{array}$$

The push out yields the abelian group $\Gamma_{\mathbb{Z}/3}(E_A)$ and we have the Yoneda equivalence of extensions

$$\begin{array}{llllllll} \mathcal{L}: & LA & \rightarrowtail & \otimes^3 A & \xrightarrow{\delta} & \tilde{\Gamma}(E_A) & \twoheadrightarrow & \Gamma A \\ & \| & & \downarrow [[1,1],1] & & \downarrow q & & \| \\ \mathcal{L}': & LA & \xrightarrow{\cdot 3} & LA & \xrightarrow{\delta'} & \Gamma_{\mathbb{Z}/3}(E_A) & \twoheadrightarrow & \Gamma A \end{array}$$

We consider the natural exact sequence

$$\Gamma_{\mathbb{Z}/3} : \ 0 \to \mathbb{Z}/3 \otimes LA \to \Gamma_{\mathbb{Z}/3}(E_A) \to \Gamma A \to 0$$

as an element

$$\Gamma_{\mathbb{Z}/3} \in Ext^1_{\mathbf{nil}}(\Gamma, \mathbb{Z}/3 \otimes L) \tag{4.3.2}$$

The short exact sequence

$$0 \to LA \xrightarrow{\cdot 3} LA \xrightarrow{p} \mathbb{Z}/3 \otimes LA \to 0$$

induces in the ususal way a long exact sequence of Ext-groups:

$$Ext^1_{\mathbf{nil}}(\Gamma, L) \xrightarrow{p_*} Ext^1_{\mathbf{nil}}(\Gamma, \mathbb{Z}/3 \otimes L) \xrightarrow{\beta} Ext^2_{\mathbf{nil}}(\Gamma, L) \xrightarrow{\cdot 3} Ext^2_{\mathbf{nil}}(\Gamma, L)$$

According to Hilton and Stammbach on IV.9.3 p.153 in [HS] we know for the Bockstein homomorphism β that

$$\beta(\Gamma_{\mathbb{Z}/3}) = < \mathcal{L}' > = < \mathcal{L} > . \tag{4.3.3}$$

Hence we get $< \mathcal{L} > = 0$ if $\Gamma_{\mathbb{Z}/3} \in image(p_*)$. In (4.4) we define a functor $\Gamma_{\mathbb{Z}}$ together with a commutative diagram of extensions

$$\begin{array}{llllllllll} \Gamma_{\mathbb{Z}/3}: & 0 & \to & \mathbb{Z}/3 \otimes L(A) & \to & \Gamma_{\mathbb{Z}/3}(E_A) & \to & \Gamma A & \to & 0 \\ & & & \uparrow p & & \uparrow \bar{p} & & \| & & \\ \Gamma_{\mathbb{Z}}: & 0 & \to & L(A) & \to & \Gamma_{\mathbb{Z}}(E_A) & \to & \Gamma A & \to & 0 \end{array}$$

This implies $p_*\Gamma_{\mathbb{Z}} = \Gamma_{\mathbb{Z}/3}$ and hence $< \mathcal{L} >= 0$. In fact in the next section we prove that one has isomorphisms

$$(4.3.4)\qquad \begin{array}{ccc} Ext^1_{\mathbf{nil}}(\Gamma, L) & \xrightarrow{p_*} & Ext^1_{\mathbf{nil}}(\Gamma, \mathbb{Z}/3 \otimes L) \\ \| & & \| \\ \mathbb{Z} & \xrightarrow{p} & \mathbb{Z}/3 \end{array}$$

where $\Gamma_{\mathbb{Z}}$ and $\Gamma_{\mathbb{Z}/3}$ are generators corresponding to $1 \in \mathbb{Z}$ and $1 \in \mathbb{Z}/3$ respectively. This implies by the Bockstein operator β above.

4.3.5. PROPOSITION: *$Ext^2_{\mathbf{nil}}(\Gamma, L)$ has no 3-torsion.*

In the same way we see that $Ext^2_{\mathbf{nil}}(\Gamma, L)$ has no odd torsion. In fact, Pirashvili shows in the appendix of this book that $Ext^2_{\mathbf{nil}}(\Gamma, L) = 0$ is the trivial group.

4.4. Computation of Ext-groups

Let R be a commutative ring with $1 \in R$. We shall prove the following result for the functors

$$\Gamma,\ R \otimes L : \mathbf{nil} \to \mathbf{ab} \to \mathbf{Ab}$$

which carry E_A to $\Gamma(A)$ and $R \otimes_{\mathbb{Z}} L(A, 1)_3$ respectively.

4.4.1. THEOREM: *Suppose R has no 2-torsion. Then there is an isomorphism of abelian groups*

$$Ext^1_{\mathbf{nil}}(\Gamma, R \otimes L) = R$$

which is natural for homomorphism $\varphi : R \to R'$ between rings; that is the following diagram commutes.

$$\begin{array}{ccc} Ext^1_{\mathbf{nil}}(\Gamma, R \otimes L) & = & R \\ \downarrow{\scriptstyle (\varphi\otimes 1)_*} & & \downarrow{\scriptstyle \varphi} \\ Ext^1_{\mathbf{nil}}(\Gamma, R' \otimes L) & = & R' \end{array}$$

We prove this result by lemma (4.4.4) and lemma (4.4.6) below. The theorem is also a consequence of results of Pirashvili in the appendix below. The advantage of our proof is the fact that we are able to describe explicit functors $\lambda\Gamma_R$ representing the elements of $Ext^1_{\mathbf{nil}}$. In order to construct elements in $Ext^1_{\mathbf{nil}}(\Gamma, R \otimes L)$ we use a universal construction as follows.

4.4.2. DEFINITION: Let $\lambda \in R$ and let $G = (G, +, -, 0)$ be a group. (G need not be abelian though we write the group law additively.) Recall that we write $L(A) = L(A, 1)_3$. We consider diagrams

$$G \xrightarrow{f} A \xleftarrow{F} R \otimes L(G^{ab})$$

with the following properties (1), (2), (3). We call (f, F) a (G, λ)-*pair*.

(1) The function F is a homomorphism between abelian groups. We write for $x, y, z \in G$
$\Delta_\lambda(x, y, z) = F(2\lambda \otimes [[\bar{x}, \bar{y}], \bar{z}] + \lambda \otimes [[\bar{z}, \bar{x}], \bar{y}])$
where $\bar{x} \in G^{ab}$ is represented by x.

(2) The function f satisfies $F(-x) = f(x)$.

(3) The cross effect $[x, y] = f(x + y) - f(x) - f(y)$ of f satisfies
$[x + y, z] = [x, y] + [y, z] + \Delta_\lambda(x, y, z)$,
$[x, y + z] = [x, y] + [x, z] + \Delta_\lambda(x, y, z)$.

Let

$$G \xrightarrow{\tilde{\gamma}} \lambda\Gamma_R(G) \xleftarrow{i} R \otimes L(G^{ab}) \tag{4.4.3}$$

be the *universal* (G, λ)-*pair*. That is, for every (G, λ)-pair (f, F) as in (4.4.2) there is a unique homomorphism $(f, \lambda F)^\square$ for which the following diagram commutes

$$\begin{array}{ccccc} G & \xrightarrow{\tilde{\gamma}} & \lambda\Gamma_R(G) & \xleftarrow{i} & R \otimes L(G^{ab}) \\ & {}_f\searrow & \downarrow (f,F)^{\square} & \swarrow_F & \\ & & A & & \end{array}$$

For $\lambda = 1 \in R$ we write $\lambda\Gamma_R(G) = \Gamma_R(G)$.

4.4.4. LEMMA: *Let R be a ring with no 2-torsion. For a free nil(2)-group E_A the universal (E_A, λ)-pair yields a short exact sequence in* **Ab**

$$\lambda\Gamma_R : 0 \to R \otimes L(A) \xrightarrow{i} \lambda\Gamma_R(E_A) \xrightarrow{p} \Gamma(A) \to 0$$

which is natural in $E_A \in$ **nil**. *The map $R \to Ext^1_{\mathbf{nil}}(\Gamma, R \otimes L)$ in (4.4.1) carries $\lambda \in R$ to the extension $\lambda\Gamma_R$.*

We have the following example for $R = \mathbb{Z}/3$ and $\lambda = 1$.

4.4.5. LEMMA: *For $R = \mathbb{Z}/3$ and $\lambda = 1$ the universal $(E_A, 1)$-pair $\lambda\Gamma_R(E_A)$ coincides with $\Gamma_{\mathbb{Z}/3}$ in (4.3.2).*

Proof: Using the definition of $\tilde{\Gamma}(E_A)$ in (4.1.1) we see that the push out of

$$\mathbb{Z}/3 \otimes LA \leftarrow ker(\tilde{p}) \to \tilde{\Gamma}(E_A)$$

has the universal property of $\lambda\Gamma_{\mathbb{Z}/3}(E_A)$ with $\lambda = 1$. Here we use the equations

$$\begin{aligned} \Delta(x,y,z) &= -[[x,z],y] - 2[[z,y],x] \\ &= [[y,x],z] + [[z,y],x] - 2[[z,y],x] \\ &= [[y,x],z] + [[x,z],y] + [[y,x],z] \\ &= 2[[x,y],z] + [[z,x],y]. \quad \square \end{aligned}$$

Proof (of (4.4.4)): We define $p : \lambda\Gamma_R(E_A) \to \Gamma(A)$ by the universal property, that is

$$p = (0, \gamma ab)^{\square})$$

where $\gamma ab : E_A \to A \to \Gamma(A)$ is defined by γ in (1.1.1) and 0 is the zero homomorphism. One readily checks that $(0, \gamma ab)$ is an (E_A, λ)-pair. By definition of p we have $pi = 0$. Moreover p is surjective since $p\tilde{\gamma} = \gamma ab$. We now choose a basis Z of the free abelian group A which is also a basis of the free nil(2)-group E_A. Moreover we choose an ordering $<$ on Z. Then the elements $\gamma(z)$, $[z,z']$, $z<z'$, $z,z'\in Z$, form a basis of ΓA. We define for p a splitting homomorphism

$$(1)\qquad \begin{cases} s:\Gamma A\to \lambda\Gamma_R(E_A) \text{ with}\\ s\gamma z=\tilde{\gamma}(z),\\ s[z,z']=[z,z'] \text{ for } z<z'. \end{cases}$$

Below we show that there is a function

$$(2)\qquad \begin{cases} g:E_A\to R\otimes L(A) \text{ with}\\ ig(x)=\tilde{\gamma}(x)-s\gamma\{x\} \text{ for } x\in E_A. \end{cases}$$

This implies that $image(i) = kernel(p)$ since $\lambda\Gamma_R(E_A)$ is generated as an abelian group by the elements $\tilde{\gamma}(x)$, $x\in E_A$, and $i(y)$, $y\in R\otimes L(A)$. The function g can actually be computed by the universal equations in (4.4.2):

Recall that we have the central extension of groups

$$\Lambda^2 A \overset{j}{\rightarrowtail} E_A \overset{ab}{\twoheadrightarrow} A$$

where j is the commutator map, that is $j(\bar{x}\wedge\bar{y})=-x-y+x+y$. Hence we can represent each element $x\in E_A$ uniquely as a finite ordered sum

$$x=(\sum_{y\in Z}^{<} n_y y)+j(\sum_{\substack{x<y\\ x,y\in Z}} a_{x,y}x\wedge y)$$

Using the distributivity laws in (4.4.2) we obtain a sum expression for $\tilde{\gamma}(x)$ which contains $s\gamma(ab(x))$ as a subsum where $ab(x)\in A$. The difference $\tilde{\gamma}(x)-s\gamma(x)$ yields canonically an element in $R\otimes L(A)$ which defines the function g above. This way we get an explicit formula for $g(x)$. We then show that the functors

$$(3)\qquad E_A \overset{(g,\gamma ab)}{\longrightarrow} R\otimes L(A)\oplus\Gamma A \overset{i_1}{\longleftarrow} R\otimes L(A)$$

form an (E_A,λ)-pair. Hence we have the commutative diagram

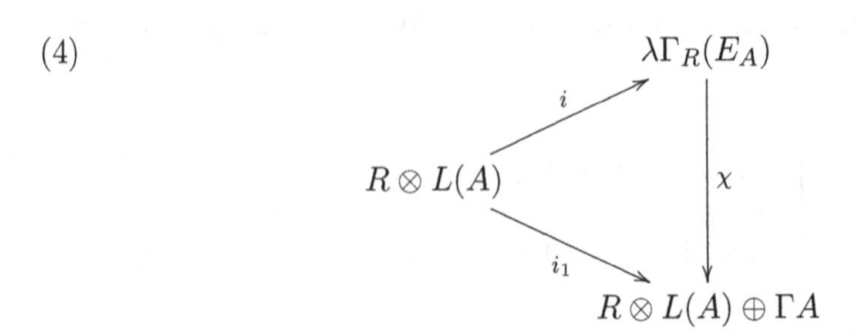

where $\chi = (i_1, (g, \gamma ab))^\square$ is defined by (3). Since i_1 is injective i is also injective. This completes the proof of lemma (4.4.4). We leave the tremendous formulas needed in this proof to the reader. The computation of g above is tedious but elementary. $\square$

4.4.6. LEMMA: *Let R be a ring with no 2-torsion. Moreover let*

$$\tau \in Ext^1_{\mathbf{nil}}(\Gamma, R \otimes L)$$

be represented by the natural exact sequence

$$R \otimes L(A) \overset{i}{\rightarrowtail} T(E_A) \overset{p}{\twoheadrightarrow} \Gamma(A)$$

*with $E_A \in$ **nil**. Then there are a canonical function $f : E_A \to T(E_A)$ and a unique $\lambda \in R$ such that (i, f) is a (G, λ)-pair.*

Proof: We denote the functor $R \otimes L$ by L_R. For $E_\mathbb{Z} = \mathbb{Z}$ we have $L_R(\mathbb{Z}) = 0$. Therefore we get the isomorphism

$$p : T(E_\mathbb{Z}) \cong \Gamma(\mathbb{Z}) \tag{1}$$

where $\gamma(1)$ is the generator of $\Gamma(\mathbb{Z}) = \mathbb{Z}$. Now we are able to define the function

$$\begin{cases} f : E_A \to T(E_A), \\ f(x) = T(\bar{x})(\gamma(1)) \end{cases} \tag{2}$$

where $\bar{x} : \mathbb{Z} \to E_A$ carries 1 to $x \in E_A$. Clearly we have

$$(3) \qquad \begin{aligned} pf(x) &= pT(\bar{x})(\gamma(1)) \\ &= \Gamma(\bar{x})p(\gamma(1)) \\ &= \Gamma(\bar{x})\gamma(1) \;=\; \gamma(\bar{x}(1)) \;=\; \gamma(x) \end{aligned}$$

so that $pf = \gamma : E_A \to A \to \Gamma A$. Moreover f is *natural* for morphisms $h : E_A \to E_B$ in **nil**, in fact

$$(4) \qquad \begin{aligned} T(h)f(x) &= T(h)T(\bar{x})(\gamma(1)) \\ &= T(h\bar{x})(\gamma(1)) \\ &= T(\overline{hx})(\gamma(1)) \\ &= f(hx). \end{aligned}$$

Now we get a function $\delta : E_A \to L_R(A)$ with

$$f(x) = f(-x) + \delta(x)$$

since $pf = \gamma$ and $\gamma(x) = \gamma(-x)$. Since f is natural for $h : E_A \to E_B$ δ also is natural, in fact

$$\begin{aligned} T(h)f(x) &= T(h)f(-x) + L_R(h)\delta(x) \text{ with} \\ T(h)f(x) &= fh(x) \text{ and} \\ fh(x) &= f(-hx) + \delta(hx). \end{aligned}$$

Hence we get $L_R(h)\delta(x) = \delta(hx)$. Now $\bar{x} : \mathbb{Z} \to E_A$ shows

$$\delta(x) = \delta(\bar{x}(1)) = L_R(\bar{x})\delta(1) = 0$$

since $\delta(1) \in L_R(\mathbb{Z}) = 0$. Therefore we actually get

$$(5) \qquad f(x) = f(-x) \text{ for } x \in E_A.$$

Next we define the bracket in $T(E_A)$ by

$$(6) \qquad [x, y] = f(x + y) - f(x) - f(y).$$

Clearly $p[x, y] = \gamma(x + y) - \gamma x - \gamma y$ is the Whitehead product in $\Gamma(A)$ which is bilinear in x, y. Therefore we can define functions

$$\Delta_1, \Delta_2 : E_A \times E_A \times E_A \to L_R(A)$$

by the equations

$$\begin{aligned} [x+y, z] &= [x, z] + [y, z] + \Delta_1(x, y, z), \\ [x, y+z] &= [x, y] + [x, z] + \Delta_2(x, y, z). \end{aligned} \tag{7}$$

Since f is natural the functions Δ_1, Δ_2 also are natural. Clearly $[0, y] = 0 = [x, 0]$ so that $\Delta_i(x, y, z) = 0$ if x, y or z is trivial. Now let a, b, c be the generators of $E_3 = E_{\mathbb{Z}\oplus\mathbb{Z}\oplus\mathbb{Z}}$ and let $h : E_3 \to E_A$ be given by $h(a) = x, h(b) = y, h(c) = z \in E_A$. Then

$$\begin{aligned} \Delta_i(x, y, z) &= \Delta_i(ha, hb, hc) \\ &= L_R(h)\Delta_i(a, b, c) \end{aligned} \tag{8}$$

where $\Delta_i(a, b, c) \in L_R(\mathbb{Z}^3)$. We claim

$$\Delta_i(a, b, c) = n_i[[a, b], c] + m_i[[c, a], b] \tag{9}$$

with $n_i, m_i \in R$. In fact $L_R(\mathbb{Z}^3)$ is a free R-module generated by triple basic commutators of the form $[[u, v], w]$ with $u, v, w \in \{a, b, c\}$ with $u < v$ and $u \leq w$. We now consider for $E_2 = E_{\mathbb{Z}\oplus\mathbb{Z}}$ a homomorphism

$$h : E_3 \to E_2$$

such that ha, hb is a basis in E_2^{ab} and $hc = 0$. Then we have

$$L_R(h)\Delta_i(a, b, c) = \Delta_i(ha, hb, hc) = 0$$

so that $\Delta_i(a, b, c)$ has only trivial coordinates for basic commutators with $\{u, v, w\} = \{a, b\}$. This implies (9). Now (9) and (8) show that actually Δ_i in (7) is of the form

$$\Delta_i(x, y, z) = n_i[[x, y], z] + m_i[[z, x], y] \tag{10}$$

for $x, y, z \in E_A$. In particular we get

$$\Delta_i(x, x, x) = 0 \tag{11}$$

and Δ_i is linear in each variable x, y, z. We now have the rules

$$\begin{aligned} 0 = [0, y] = [-x + x, y] = [-x, y] + [x, y] + \Delta_1(-x, x, y), \\ 0 = [x, 0] = [x, -y + y] = [x, -y] + [x, y] + \Delta_2(x, -y, y) \end{aligned}$$

so that

$$\left\{ \begin{array}{lcl} [-x, y] & = & -[x, y] + \Delta_1(x, x, y) \\ [x, -y] & = & -[x, y] + \Delta_2(x, y, y). \end{array} \right. \tag{12}$$

Moreover by (5) we get for $x, y \in E_A$

$$\begin{array}{lcl} f(x + y) & = & f(x) + f(y) + [x, y], \\ f(-(x + y)) & = & f(-y - x) = f(-y) + f(-x) + [-y, -x] \\ & = & f(y) + f(x) + [-y, -x], \end{array}$$

where

$$\begin{array}{lcl} [-y, -x] & = & -[y, -x] + \Delta_1(y, y, -x) \\ & = & -(-[y, x] + \Delta_2(y, x, x)) - \Delta_1(y, y, x) \\ & = & [y, x] - \Delta_2(y, x, x) - \Delta_1(y, y, x), \end{array}$$

so that

$$[x, y] = [y, x] - \Delta_2(y, x, x) - \Delta_1(y, y, x). \tag{13}$$

If we replace x in(13) by $x + x'$ we thus get the equations

$$\left\{ \begin{array}{lcl} [x + x', y] & = & [y, x + x'] - \Delta_2(y, x + x', x + x') - \Delta_1(y, y, x + x'), \\ [x, y] & = & [y, x] - \Delta_2(y, x, x) - \Delta_1(y, y, x), \\ [x', y] & = & [y, x'] - \Delta_2(y, x', x') - \Delta_1(y, y, x'). \end{array} \right.$$

Using (7) we thus get

$$\begin{aligned} & \Delta_1(x,x',y) - \Delta_2(y,x,x) - \Delta_1(y,y,x) - \Delta_2(y,x',x') - \Delta_1(y,y,x') \\ = \ & \Delta_2(y,x,x') - \Delta_2(y,x+x',x+x') - \Delta_1(y,y,x+x') \end{aligned}$$

Hence we get

$$\Delta_1(x,x',y) = \Delta_2(y,x,x') - \Delta_2(y,x,x') - \Delta_2(y,x',x).$$

This implies

$$\Delta_1(x,x',y) = -\Delta_2(y,x',x) \tag{14}$$

or equivalently

$$n_1[[x,x'],y] + m_1[[y,x],x'] = -n_2[[y,x',x] - m_2[[x,y],x']$$

where $[[y,x',x] = -[[x,y],x'] - [[x',x],y]$. This implies

$$m_1 = n_2 - m_2 \text{ and } n_1 = n_2. \tag{15}$$

On the other hand we have

$$0 = f(0) = f(-x+x) = f(-x)+f(x)+[-x,x] = 2f(x)-[x,x] \text{ (see (12), (11))}$$

so that

$$[x,x] = 2f(x) \text{ for } x \in E_A. \tag{16}$$

Next we get for a commutator

$$\begin{aligned} & f(-(y+x)+(x+y)) = f(-(y+x)) + f(x+y) + [-(y+x),x+y] \\ = \ & f(y+x) + f(x+y) - [y+x,x+y] + \Delta_1(y+x,y+x,x+y) \\ = \ & f(y+x) + f(x+y) - [y+x,x+y] && \text{(see} \\ = \ & 2f(y) + 2f(x) + [y,x] + [x,y] - [y+x,x+y] \\ = \ & [y,y] + [x,x] + [y,x] + [x,y] - [y+x,x+y] && \text{(see} \end{aligned}$$

$$\begin{aligned}[y+x,x+y] &= [y,x+y]+[x,x+y]+\Delta_1(y,x,x+y)\\ &= [y,x]+[y,y]+\Delta_2(y,x,y)+[x,x]+[x,y]\\ &\quad + \Delta_2(x,x,y)+\Delta_1(y,x,x+y).\end{aligned}$$

Hence we get

$$f(-x-y+x+y) = -\Delta_2(x+y,x,y)-\Delta_1(y,x,x+y). \tag{17}$$

Thus we get the folloing equations where $m_1 = n_2 - m_2$ by (15).

$$\begin{aligned}f(-x-y+x+y) &= -\ n_2[[x+y,x],y]-m_2[[y,x+y],x]\\ &\quad -\ n_1[[y,x],x+y]-m_1[[x+y,y],x]\\ &= -\ n_2[[x+y,x],y]-n_2[[y,x+y],x]\\ &\quad -\ n_1[[y,x],x+y]\\ &= [[x,y],x](-n_2-n_1+2n_2)\\ &\quad +\ [[y,x],y](-n_2+2n_2-n_1)\\ &= (n_2-n_1)([[x,y],x]+[[y,x],y])\\ &= 0 \text{ by (15)}.\end{aligned}$$

Hence we have shown

$$f(-x-y+x+y)=0. \tag{18}$$

Next we compute for the commutator $(x,y) = -x-y+x+y$:

$$\begin{aligned}f(z+(x,y)) &= f(z)+f(x,y)+[z,(x,y)]\\ &= f(z)+[z,(x,y)] \text{ by (17) where}\end{aligned}$$

$$\begin{aligned}[z,(x,y)] &= [z,-(x+y)+(x+y)]\\ &= [z,-(y+x)]+[z,x+y]+\Delta_2(z,-(y+x),x+y)\end{aligned}$$

where $[z,-(x+y)] = -[z,y+x]+\Delta_2(z,y+x,y+x)$, hence

$$\begin{aligned}[z,(x,y)] &= -[z,y+x]+[z,x+y]\\ &= -\Delta_2(z,y,x)+\Delta_2(z,x,y).\end{aligned}$$

This implies

$$f(z + (x, y)) = f(z) - \Delta_2(z, y, x) + \Delta_2(z, x, y). \tag{19}$$

On the other hand we have $x + y - y - x + y + x = x + y$, so that

$$\begin{aligned} f(x + y + (x, y)) &= f(y + x) = f(y) + f(x) + [y, x] \\ &= f(x) + f(y) + [x, y] - \Delta_2(x + y, x, y) + \Delta_2(x + y, y, x) \end{aligned}$$

where

$$[x, y] = [y, x] - \Delta_2(y, x, x) - \Delta_1(y, y, x) \text{ by } (13).$$

Hence we get

$$0 = \Delta_2(x + y, y, x) - \Delta_2(x + y, x, y) - \Delta_2(y, x, x) - \Delta_1(y, y, x)$$

where by (14) we have $-\Delta_1(y, y, x) = \Delta_2(x, y, y)$. Hence we get

$$\begin{aligned} n_2[[x + y, y], x] + m_2[[x, x + y], y] \\ - n_2[[x + y, x], y] - m_2[[y, x + y], x] \\ - n_2[[y, x], x] - m_2[[x, y], x] \\ + n_2[[x, y], y] + m_2[[y, x], y] = 0 \end{aligned}$$

This implies

$$[[x, y], x](n_2 - 2m_2 + 2n_2 - m_2 - n_2 - m_2) = 0$$

and hence $2n_2 - 4m_2 = 0$. On the other hand

$$[[y, x], y](-2n_2 + m_2 - n_2 + 2m_2 + n_2 + m_2) = 0$$

and hence $-2n_2 + 4m_2 = 0$. Thus we have shown

$$2n_2 = 4m_2 \tag{20}$$

where $n_2 = m_1 + m_2$ and $n_1 = n_2$ by (15). Hence we get $2m_1 = 2m_2$. Since R has no 2-torsion we have $m_1 = m_2$. Hence we get for $m_1 = m_2 = \lambda \in R$

$$\Delta_1 = \Delta_2 = \Delta_\lambda(x,y,z) = 2\lambda[[x,y],z] + \lambda[[z,x],y].$$

This completes the proof of (4.4.5). □

4.5. The splitting of TL and a model of $\mathbf{CW}(2,4)/D_\Gamma$

We have for each free nil(2)-group E_A the commutative diagram with $L(A) = L(A,1)_3$,

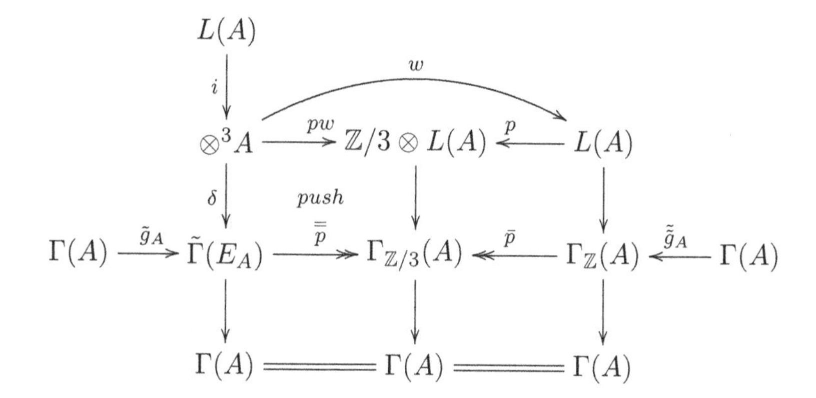

Here $L(A) = L(A,1)_3$ and $W = [[1,1],1]$. The columns of the diagram are exact; compare (4.3.1) and (4.3.3).

4.5.1. LEMMA: *Let $H, H' \in \otimes^3 A$. Then $\delta(H) = \delta(H')$ and $W(H) = W(H')$ imply $H = H'$.*

Proof: $\delta(H) = \delta(H')$ implies that there is $\alpha \in L(A)$ with $H' = H + i(\alpha)$. Then $W(H) = W(H')$ implies $Wi(\alpha) = 3\alpha = 0$. This implies $\alpha = 0$ since $L(A)$ is free abelian. □

We choose for $\tilde{g}_A$ in (4.1.6) a homomorphism $\approx{g}_A$ as in the diagram above with

(4.5.2) $$\bar{p}\,\overset{\approx}{g}_A = \overset{=}{p}\,\tilde{g}_A.$$

Then such a choice defines a splitting function for the linear extension **TL** in

(4.1.6)

$$s : \mathbf{E}(2,4) \to \mathbf{TL} \tag{4.5.3}$$

as follows. For $(\xi,\eta) : f \to g$ in $\mathbf{E}(2,4)$ with $f : B' \to \Gamma(A')$ and $g : B \to \Gamma(A)$ let $s(\xi,\eta) = (\xi,\eta,H(\xi,\eta))$ be defined by the unique element $H(\xi,\eta) \in Hom(B',\otimes^3 A)$ with

$$\begin{cases} \delta H(\xi,\eta) &= -\tilde{\Gamma}(\eta)\tilde{f}_A f + \tilde{g}_A g\xi^{ab}, \\ WH(\xi,\eta) &= -\Gamma_{\mathbb{Z}}(\eta)\, \tilde{\tilde{f}}_A\, f + \tilde{\tilde{g}}_A\, g\xi^{ab}. \end{cases}$$

Here $H(\xi,\eta)$ is well defined by (4.5.1). For this we observe that the square labelled *push* in the diagram above is also a pull back of abelian groups.

We now study the category $\mathbf{CW}(2,4)/D_\Gamma$ in (4.1.9). We have a commutative diagram of linear extensions

$$\begin{array}{ccccc} Hom(\_,L) & \longrightarrow & \mathbf{LT} & \longrightarrow & \mathbf{E}(2,4) \\ \downarrow & & \downarrow & & \downarrow \\ L & \longrightarrow & \mathbf{CW}(2,4)/D_\Gamma & \longrightarrow & \mathbf{H}(2,4) \end{array} \tag{4.5.4}$$

Here the natural system $L = D/D_\Gamma = D/qE_\Gamma$ is defined by (4.1.8). According to (1.3.3) we get $L(\xi,\eta)$ by the cokernel:

$$L(\xi,\eta) = cok(d : Hom(A', cok(g)) \to Hom(B', L(g)))$$

where $d(\alpha) = q(\alpha \otimes \eta)\tau f$ is the composite

$$B' \xrightarrow{f} \Gamma(A') \xrightarrow{\tau} A' \otimes A' \xrightarrow{\alpha\otimes\eta} cok(g) \otimes A \xrightarrow{q} L(g).$$

Here $L(g)$ is defined by the push out

$$\begin{array}{ccc} \Gamma(A) \otimes A & \xrightarrow{W} & L(A,1)_3 \\ \downarrow & & \downarrow \\ cok(g) \otimes A & \xrightarrow{q} & L(g) \end{array}$$

with W defined by $W(\gamma(x) \otimes g) = -[[y,x],x]$ as in (2.2.7). As in (2.6.5) we obtain a map

$$\psi : Hom(A', \Lambda^2 A) \to L(\xi, \eta) \tag{4.5.5}$$

which carries $\alpha : A' \to \Lambda^2 A$ to the element in $L(\xi, \eta)$ represented by the composite

$$B' \xrightarrow{f} \Gamma(A') \xrightarrow{\tau} A' \otimes A' \xrightarrow{H \otimes \eta - \eta \otimes A} A \otimes A \otimes A \xrightarrow{[[1,1],1]} L(A,1)_3 \longrightarrow L(g).$$

Here $H : A' \to A \otimes A$ is a lift of α via the projection $A \otimes A \to \Lambda^2 A$. As in (2.6.5) one can check that ψ is well defined.

The category $\mathbf{CW}(2,4)/D_\Gamma$ now is completely algebraically described by the following result. Consider the pull back category

$$\begin{array}{ccc} \tilde{\mathbf{H}}(2,4) & \longrightarrow & \mathbf{nil} \\ \downarrow & & \downarrow \\ \mathbf{H}(2,4) & \xrightarrow[p_1]{} & \mathbf{ab} \end{array}$$

where $p_1(\xi, \eta) = \eta$. Then the section s in (4.5.3) is obtained as a composite $\mathbf{E}(2,4) \to \tilde{\mathbf{H}}(2,4) \to \mathbf{TL}$.

4.5.6. THEOREM: *The composite functor*

$$\bar{s} : \tilde{\mathbf{H}}(2,4) \xrightarrow{s} \mathbf{TL} \to \mathbf{CW}(2,4)/D_\Gamma$$

yields a push forward of linear extensions

$$\begin{array}{ccc} Hom(\_, \Lambda^2) & \xrightarrow{\psi} & L \\ \downarrow & & \downarrow \\ \tilde{\mathbf{H}}(2,4) & \xrightarrow{\bar{s}} & \mathbf{CW}(2,4)/D_\Gamma \\ \downarrow & & \downarrow \\ \mathbf{H}(2,4) & = & \mathbf{H}(2,4) \end{array}$$

4.5.7. Corollary: *The cohomology class represented by* $\mathbf{CW}(2,4)/D_\Gamma$ *is given by*

$$\{\mathbf{CW}(2,4)/D_\Gamma\} = \psi_* p_1^* \{\mathbf{nil}\}.$$

Hence this class is an element of order at most 2.

Proof (of (4.5.6)): Definition (4.1.10) implies that the functor $\bar{s}$ is ψ-equivariant. $\square$

CHAPTER 5

THE CATEGORY $\mathbf{T}\Gamma$ AND AN ALGEBRAIC MODEL OF $\mathbf{CW}(2,4)$

We have seen in chapter 3 that the category $\mathbf{CW}(2,4)$ can be described in terms of a category $\mathbf{T}$ which is a pull back of

$$\mathbf{TL} \to \mathbf{E}(2,4) \leftarrow \mathbf{T}\Gamma.$$

The category $\mathbf{TL}$ was computed in chapter 4, in fact $\mathbf{TL}$ is a split extension over $\mathbf{E}(2,4)$. We now describe the category $\mathbf{T}\Gamma$ algebraically and we then obtain an algebraic category equivalent to $\mathbf{CW}(2,4)$.

5.1. The quadratic refinement $\bar{\Gamma}$

We introduce a functor $\bar{\Gamma}$ which is a further refinement of Whitehead's quadratic functor Γ. The functor $\bar{\Gamma}$ is defined (similarly to $\tilde{\Gamma}$ in (4.1.1)) by a universal property with respect to certain distributivity laws which, however, depend on the choice of a "symmetric square map" θ. Using this quadratic refinement $\bar{\Gamma}$ we describe an algebraic category $\Gamma\mathbf{T}$ which is equivalent to the topological track category $\mathbf{T}\Gamma = \mathbf{T}(2,4)/E_\Gamma$. The results in this section are proved in chapter 7 by use of theory of quadratic chain complexes, see [BCH].

For an abelian group A we get the n-th *symmetric power* $SP_n(A)$ by the quotient group

$$SP_n(A) = \otimes^n A/S(A) \tag{5.1.1}$$

where $S(A)$ is the subgroup generated by the relations $(x_1, \ldots, x_n \in A)$

$$x_1 \otimes \cdots \otimes x_n - x_{\sigma(1)} \otimes \cdots \otimes x_{\sigma(n)} \in S(A) \tag{1}$$

where σ is a permutation. If $A = A_2$ is a $\mathbb{Z}/2$-vector space we get by (1.1.2) the equation

$$A_2 \hat{\otimes} A_2 = \hat{\otimes}^2(A_2) = SP_2(A_2). \tag{2}$$

We also need a functor $\tilde{\otimes}^4$ which carries abelian groups to abelian groups, we define this functor by the quotient group

$$\tilde{\otimes}^4(A) = (\otimes^4 A)/R(A) \tag{5.1.2}$$

Here $R(A)$ is the subgroup generated by the relations (1), (2) and (3) with $x, y, z, x', y' \in A$.

$$x \otimes y \otimes x' \otimes y' + x' \otimes y' \otimes x \otimes y \ \in R(A), \tag{1}$$

$$x \otimes y \otimes x \otimes z - z \otimes y \otimes z \otimes x \ \in R(A), \tag{2}$$

$$x \otimes y \otimes z \otimes y - y \otimes x \otimes z \otimes x \ \in R(A). \tag{3}$$

The relations (2) and (3) imply the relations (4) and (5) respectively:

$$x \otimes y \otimes x' \otimes z + x' \otimes y \otimes x \otimes z \ \in R(A), \tag{4}$$

$$x \otimes y \otimes z \otimes y' + x \otimes y' \otimes z \otimes y \ \in R(A). \tag{5}$$

This follows by replacing x in (2), resp. y in (3), by $x + x'$, resp. $y + y'$. We are mainly interested in the case when $A = A_2$ is a $\mathbb{Z}/2$-vector space. In this case (1), (4) and (5) are special relations of the form (5.1.1)(1). We denote the equivalence class of $x \otimes y \otimes x' \otimes y'$ in $\tilde{\otimes}^4 A$ by $xyx'y'$. The relations (4) and (5) show that one has a natural transformation

$$\begin{cases} T : (\hat{\otimes}^2 A) \otimes (\hat{\otimes}^2 A) & \twoheadrightarrow & \tilde{\otimes}^4 A, \\ T(x \hat{\otimes} y) \otimes (x' \hat{\otimes} y') & = & xx'yy'. \end{cases} \tag{5.1.3}$$

Moreover, relation (1) yields the natural transformation

$$(5.1.4) \qquad \begin{cases} I : (\otimes^2 A)\hat{\otimes}(\otimes^2 A) & \twoheadrightarrow & \tilde{\otimes}^4 A, \\ I(x \otimes y)\hat{\otimes}(x' \otimes y') & = & xyx'y'. \end{cases}$$

Using the transformation τ in (1.1.2) we get the natural composition

$$(1) \qquad i : (\Gamma A) \otimes \mathbb{Z}/2) \overset{\tau\otimes\mathbb{Z}/2}{\to} (\otimes^2 A) \otimes \mathbb{Z}/2 \overset{\tau}{\to} (\otimes^2 A)\hat{\otimes}(\otimes^2 A) \overset{I}{\to} \tilde{\otimes}^4 A$$

which also well is determined by the equation

$$(2) \qquad i((\gamma a) \otimes 1) = aaaa, \;\; a \in A.$$

Here $\gamma a \in \Gamma A$ is given by the quadratic map $\gamma : A \to \Gamma A$ and $\underline{1} \in \mathbb{Z}/2$ is the generator. We will need the following crucial lemma on the transformation i in (1).

5.1.5. LEMMA: *Let A be a $\mathbb{Z}/2$-vector space. Then $i : (\Gamma A) \otimes \mathbb{Z}/2 \to \tilde{\otimes} A$ is injective.*

Proof: Let Z be a basis of A. We consider the set W of words $xyx'y'$ with $x, y, x', y' \in Z$. Let $\sim$ be the equivalence relation on W generated by the relations given by (5.1.2)(1),...,(5). Then the obvious map

$$(1) \qquad q : W/\sim \to \tilde{\otimes}^4 A$$

is injective and the image of this map is a basis of the $\mathbb{Z}/2$-vector space $\tilde{\otimes}^4 A$. A basis of $(\Gamma A) \otimes \mathbb{Z}/2$ is given by the set of elements

$$(2) \qquad (\gamma x) \otimes \underline{1}, \;\; [x, y] \otimes \underline{1}, \;\; x < y, \;\; x, y \in Z.$$

The function i in (5.1.5) carries such basis elements to

$$(3) \qquad \begin{cases} i(\gamma x) \otimes \underline{1} & = & q(xxxx), \\ i[x, y] \otimes \underline{1} & = & q(xyxy) + q(yxyx). \end{cases}$$

Here the equivalence class of $xyxy$ resp. $yxyx$, in $W/\sim$ consists of five elements

(4) $\{xyxy, yyyx, yxxx, yxyy, xxyx\}$, resp. $\{yxyx, xxxy, xyyy, xyxx, yyxy\}$.

This proves that i in (5.1.5) is injective. □

For a $\mathbb{Z}/2$-vector space A we also need the commutative diagram

(5.1.6)
$$\begin{array}{ccc} (\Gamma A)\otimes\mathbb{Z}/2 & \xrightarrow{\sigma\otimes 1} & A\otimes\mathbb{Z}/2 \\ {\scriptstyle i}\downarrow & & \downarrow{\scriptstyle i} \\ \tilde{\otimes}^4 A & \xrightarrow{\sigma} & S\tilde{P}_4 A \end{array}$$

where i on both sides is injective. The group

(1)
$$S\tilde{P}_4 A = SP_4(A)/R'(A)$$

is defined by the subgroup $R'(A)$ of $SP_4(A)$ which is generated bu the relations (5.1.2)(2) or (5.1.2)(3). The natural transformation σ in the top row is the suspension in (1.1.2) and the transformation σ in the bottom row is induced by the identity of $\otimes^4 A$. Using σ in (5.1.6) one can compute the image category of the suspension functor

(2)
$$\Sigma : \mathbf{T}(2,4) \to \mathbf{T}(3,5).$$

For a group G and $n \in \mathbb{N}$ we obtain the $(\mathbb{Z}/n)$-module

(5.1.7)
$$G_n = G^{ab} \otimes_{\mathbb{Z}} \mathbb{Z}/n.$$

Similarly we get for a homomorphism $\eta : G' \to G$ the induced map $\eta_n : G'_n \to G_n$, $\eta_n = \eta^{ab} \otimes \mathbb{Z}/n$. Let $h : G \to G_n$ be the quotient map which satisfies $\eta_n h = h\eta$. The following definition describes a symmetric analogue of the "θ-groups" considered in (7.1.1).

5.1.8. DEFINITION: A *symmetric θ-group* (G, θ) consists of a group G and a function

(1)
$$\theta : G \to G_2 \hat{\otimes} G_2$$

which satisfies for $x, y \in G$

$$\theta(x+y) = \theta(x) + \theta(y) + h(x)\hat{\otimes}h(y). \tag{2}$$

Here we use $h : G \twoheadrightarrow G_2$ in (5.1.7). For simplicity we also write $xy = h(x)\hat{\otimes}h(y)$ as a product. Let (G', θ') be a further symmetric θ-group and let $\eta : G' \to G$ be a homomorphism. Then one obtains the homomorphism

$$\begin{cases} \theta(\eta) : G' \overset{h}{\twoheadrightarrow} G'_2 \overset{\theta(\eta)}{\longrightarrow} \hat{\otimes}^2 G_2, \\ \theta(\eta) = -(\eta_2\hat{\otimes}\eta_2)\theta' + \theta\eta. \end{cases} \tag{3}$$

Let **SG** be the following category. Objects are symmetric θ-groups and morphisms $\eta : (G', \theta') \to (G, \theta)$ are homomorphisms $\eta : G' \to G$ between the underlying groups. Let $\mathbf{SG}_0$ be the subcategory of **SG** consisting of morphisms η with $\theta(\eta) = 0$.

The next lemma is an easy exercise, see (7.1.2).

5.1.9. LEMMA: *A summetric θ-group (G, θ) satisfies $\theta(-x-y+x+y) = 0$ and $\theta(4x) = 0$, $(x, y \in G)$. Therefore θ admits a unique factorization*

$$\theta : G \overset{h}{\to} G_4 \overset{\bar{\theta}}{\to} G_2\hat{\otimes}G_2$$

and thus $\theta(\eta)$ depends only on $\eta_4 : G'_4 \to G_4$.

5.1.10. EXAMPLE: Let G be a group for which G_4 is a free $(\mathbb{Z}/4)$-module with basis Z. Then one has the symmetric θ-group (G, θ) where $\theta = \bar{\theta}h$ is defined by $\bar{\theta}(x) = 0$ for $x \in Z$. This follows from the more general result in (7.2.2) by use of (5.1.9).

We are now ready for the definition of the functor $\bar{\Gamma}$ which is a further refinement of Whitehead's quadratic functor Γ.

5.1.11. DEFINITION: Let (G, θ) be a symmetric θ-group. We call a function f and a homomorphism F as in the diagram

$$G \overset{f}{\to} K \overset{F}{\leftarrow} \tilde{\otimes}^4 G_2$$

a (G, θ)-pair if the following properties are satisfied. The homomorphism F maps to the center of K. Moreover for $x, y, x', y' \in G$ let $xy = hx\hat{\otimes}hy \in \hat{\otimes}^2 A$

with $A = G_2$ as in (5.1.8)(3) and let

$$xyx'y' = hxhyhx'hy' \in \tilde{\otimes}^4 A$$

where $h : G \to G_2$ is the quotient map. Then f and F satisfy the following equations (1)...(4) where we set

$$[x, y] = -f(y) + f(x + y) - f(x).$$

$$-f(x) - f(y) + f(x) + f(y) = F(xxyy), \tag{1}$$

$$f(-x) = f(x), \tag{2}$$

$$[x + y, z] = [x, z] + [y, z] + F\Delta'(x, y, z), \tag{3}$$

$$[x, y + z] = [x, y] + [x, z] + F\Delta''(x, y, z). \tag{4}$$

Here we define Δ' and Δ'' as follows. Using T in (5.1.3) let

$$\begin{aligned}
\tilde{\Delta}(x, y, z) &= T(xy \otimes \theta z + xz \otimes \theta y + yz \otimes \theta x) + xyxz, \\
\Delta'(x, y, z) &= \tilde{\Delta}(x, y, z) + zyxx + zyyx + zzyx, \\
\Delta''(x, y, x) &= \tilde{\Delta}(x, y, z) + xxyz + xyyz + xyzz.
\end{aligned}$$

A (G, θ)-pair

$$G \xrightarrow{\gamma} \bar{\Gamma}(G, \theta) \xleftarrow{\delta} \tilde{\otimes}^4 G_2 \tag{5}$$

is *universal* if for each (G, θ)-pair (f, F) as above there is a unique homomorphism between groups

$$(f, F)^{\square} : \bar{\Gamma}(G, \theta) \to K \tag{6}$$

with $(f, F)^{\square}\gamma = f$ and $(f, F)^{\square}\delta = F$. The universal (G, θ)-quadratic pair exists and yields a functor

$$\bar{\Gamma} : \mathbf{SG}_0 \to \mathbf{Gr}. \tag{7}$$

For a homomorphism $\eta : (G', \theta') \to (G, \theta)$ in $\mathbf{SG}_0$ (with $\theta(\eta) = 0$) the induced homomorphism is given by $\bar{\Gamma}(\eta) = (\gamma\eta, \delta\tilde{\otimes}^4(\eta_2))^{\square}$.

5.1.12. DEFINITION: Let (G',θ') and (G,θ) be symmetric θ groups and let $\eta: G' \to G$ be a homomorphism, (that is, η is a morphism in **SG**). Then we define (see (5.1.3))

$$\left\{ \begin{array}{l} \Delta_\eta : G' \to \tilde{\otimes}^4 G_2 \\ \Delta_\eta(x) = T[(\eta_s \hat{\otimes} \eta_2)\theta(x) \otimes \theta(\eta)(x)], \ x \in G'. \end{array} \right. \tag{1}$$

Consider the following diagram in which the rows are universal quadratic pairs.

$$\begin{array}{ccccc} G' & \xrightarrow{\gamma'} & \bar{\Gamma}(G',\theta') & \xleftarrow{\delta'} & \tilde{\otimes}^4 G_2' \\ \downarrow{\scriptstyle \eta} & \searrow^{f} & \downarrow & \swarrow^{F} & \downarrow{\scriptstyle \tilde{\otimes}^4 \eta_2} \\ G & \xrightarrow{\gamma} & \bar{\Gamma}(G,\theta) & \xleftarrow{\delta} & \tilde{\otimes}^4 G_2 \end{array} \tag{2}$$

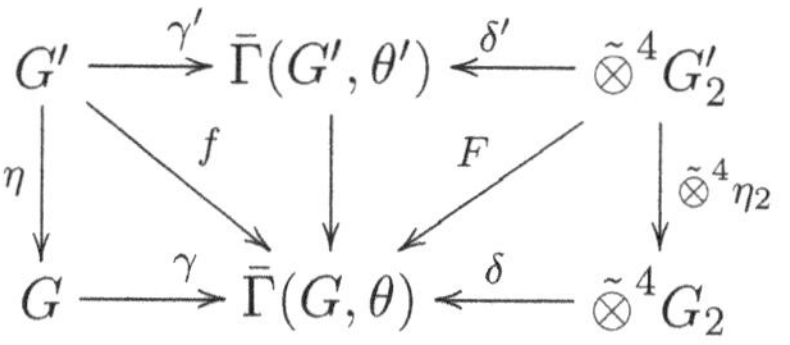

Here we set

$$F = \delta(\tilde{\otimes}^4 \eta_2), \ f = \gamma\eta + \delta\Delta_\eta. \tag{3}$$

Then one can check that (F,f) is a (G',θ')-pair so that the homomorphism

$$\bar{\Gamma}(\eta) = (f,F)^\square : \bar{\Gamma}(G',\theta') \to \bar{\Gamma}(G,\theta) \tag{4}$$

is defined for η in **SG**. We call $\bar{\Gamma}(\eta)$ the *generalized induced map* for $\bar{\Gamma}$. These generalized induced maps satisfy a composition law as in (5.1.17) below so that they do not define a functor on **SG**. This is important for the definition of track composition in the algebraic category Γ**T** below.

We now describe some properties of the universal (G,θ)-pair. We first observe that $\bar{\Gamma}(G,\theta)$ (similarly to $\tilde{\Gamma}(G)$ in (4.1) above) depends only on the quotient $G/\Gamma_3 G$ of nilpotency degree 2. For this we consider the quotient map in $\mathbf{SG}_0$

$$q : (G,\theta) \to (G/\Gamma_3 G, \theta)$$

which is well defined, by (5.1.9).

5.1.13. PROPOSITION: *The induced map*

$$q_* : \bar{\Gamma}(G,\theta) \to \bar{\Gamma}(G/\Gamma_3 G, \theta)$$

is an isomorphism.

This is a consequence of the next lemma.

5.1.14. LEMMA: *Let (f, E) be a (G, θ)-pair as in (5.1.11). Then for $x, y, z \in G$ the following formulas hold:*

$$\begin{aligned} f(-x-y+x+y) &= F(xyxy), \\ f(z-x-y+x+y) &= f(z) + F(xyxy + zxzy + xyzy). \end{aligned}$$

The proof of the lemma is a lengthy but straightforward application of the distributivity laws in (5.1.11), it also follows from (7.1.13). Our main result on the functor $\bar{\Gamma}$ is the following theorem.

5.1.15. THEOREM: *For any symmetric θ-group (G, θ) there is a sequence of homomorphisms between groups*

$$0 \to \Gamma(G_2) \otimes \mathbb{Z}/2 \xrightarrow{i} \tilde{\otimes}^4 G_2 \xrightarrow{\delta} \bar{\Gamma}(G, \theta) \xrightarrow{\bar{p}} \Gamma(G^{ab}) \to 0$$

which is natural for maps in **SG**, *see (5.1.12), and which satisfies $\delta i = 0$, $p\delta = 0$. Moreover, if G is a free group (or if G is a free nil(2)-group) and (G, θ) is given by a basis of G as in (5.1.10) then this sequence is exact.*

The natural map i in (5.1.15) is given by (5.1.4)(1) and the map δ is part of the universal property. Moreover the projection $\bar{p}$ is given by $\bar{p} = (\gamma h, 0)^{\square}$ where $\gamma h : G \to G^{ab} \to \Gamma(G^{ab})$, see (1.1.1) and (5.1.11)(6). It is a consequence of (5.1.11)(4) that $\delta i = 0$ and we get $\bar{p}\delta = 0$ by definition of $\bar{p}$. We shall prove the complicated part of theorem (5.1.15), namely the exactness for a free group G, in (7.1.3).

We now consider the *composition law for generalized induced maps.* Let

$$\eta\eta' : (G'', \theta'') \to (G', \theta') \to (G, \theta)$$

be maps in **SG**, then we define the homomorphism

$$(5.1.16) \qquad \begin{cases} \Theta(\eta, \eta') : \otimes^2 G_2'' \to \tilde{\otimes}^4 G_2, \\ \Theta(\eta, \eta') = T[(\hat{\otimes}^2 \eta_2)\theta(\eta') \otimes \theta(\eta)\eta_2'], \end{cases}$$

where we use $\theta(\eta)$ in (5.1.8)(3) and T in (5.1.3). Using the composition

$$(h \otimes h)\tau\bar{p} : \bar{\Gamma}(G'', \theta'') \to \Gamma(G'')^{ab} \to \otimes^2(G'')^{ab} \to \otimes^2 G''_2$$

with τ in (1.1.2) we get

5.1.17. LEMMA: $\bar{\Gamma}(\eta)\bar{\Gamma}(\eta') = \bar{\Gamma}(\eta\eta') + \delta\Theta(\eta, \eta')(h \otimes h)\tau\bar{p}$.

We now consider a free nil(2)-group E_A with a fixed basis $Z \subset E_A$ so that $E_A = G/\Gamma_3 G$ where $G = < Z >$ is a free group with $G^{ab} = E_A^{ab} = A$. Using the basis Z we define the symmetric square map

$$\theta : E_A \to A_2 \hat{\otimes} A_2, \ A_2 = A \otimes \mathbb{Z}/2,$$

by $\theta(x) = 0$ for $x \in Z$, see (5.1.10). For the symmetric θ-group (E_A, θ) we obtain by (5.1.15) the exact sequence

$$0 \to \Gamma(A) \otimes \mathbb{Z}/2 \xrightarrow{i} \tilde{\otimes}^4 A_2 \xrightarrow{\delta} \bar{\Gamma}(E_A, \theta) \xrightarrow{\bar{p}} \Gamma(A) \to 0 \tag{5.1.18}$$

which is natural for homomorphisms $\eta : E_{A'} \to E_A$ between free nil(2)-groups. For the naturality we use the generalized induced homomorphisms $\bar{\Gamma}(\eta)$ in (5.1.12) which have the following additional property. Let $\eta' : E_{A'} \to E_A$ be given with $(\eta')^{ab} = \eta^{ab}$, then we can choose

$$H_\eta \in T_2(\eta, \eta'), \ H_\eta : A' \to A \otimes A \tag{1}$$

as in (3.1.3)(1). Consider the diagram of homomorphisms between groups

$$\begin{array}{ccc} \bar{\Gamma}(E_{A'}, \theta') & \xrightarrow{\bar{\Gamma}(\eta), \bar{\Gamma}(\eta')} & \bar{\Gamma}(E_A, \theta) \\ \bar{p} \downarrow & & \uparrow \delta \\ \Gamma(A') & \xrightarrow{\bar{H}_\eta} & \tilde{\otimes}^4 A_2 \end{array} \tag{2}$$

Here $\bar{H}_\eta$ is defined by the sum

$$\bar{H}_\eta = m(H_\eta \otimes \eta^{ab})\tau + n(\eta^{ab} \otimes H_\eta)\tau + I(H_\eta \otimes H_\eta)\tau \tag{3}$$

where $\tau : \Gamma(A') \to A' \otimes A'$ and where m, n, I are the following natural homomorphisms:

$$\begin{aligned}
m &: \otimes^3 A \to \tilde{\otimes}^4 A_2, \ m(x \otimes y \otimes z) = xyzy, \\
n &: \otimes^3 A \to \tilde{\otimes}^4 A_2, \ n(x \otimes y \otimes z) = xyxz, \\
I &: \otimes^4 A \to \tilde{\otimes}^4 A_2, \ I(x \otimes y \otimes x' \otimes y') = xyx'y'.
\end{aligned}$$

Now lemma (5.1.14) implies the formula

$$-\bar{\Gamma}(\eta) + \bar{\Gamma}(\eta') = \delta \circ \bar{H}_\eta \circ \bar{p}. \tag{4}$$

This is the analogue of formula (4.1.5)(4). Indeed the sequence (5.1.18) plays a similar role to the exact sequence (4.1.5) for $\tilde{\Gamma}(E_A)$. As in (4.1.6) we define by the exact sequence above the following "algebraic track category".

5.1.19. DEFINITION: Let $\Gamma\mathbf{T}$ be the following algebraic category which is part of a linear extension

$$E_\Gamma + \rightarrowtail \Gamma\mathbf{T} \twoheadrightarrow \mathbf{E}(2,4). \tag{1}$$

For each object $g : B \to \Gamma(A)$ in $\mathbf{E}(2,4)$ we fix a homomorphism

$$\bar{g} : E_B \to \bar{\Gamma}(E_A, \theta)$$

with $\bar{p}\bar{g} = gh$ where $h : E_B \to B$ is the quotient map. We consider $\bar{g}$ as an object in $\Gamma\mathbf{T}$ which is identified by the functor p in (1) with the object g, so that p in this sense is the identity on objects. A morphism $\bar{f} \to \bar{g}$ in $\Gamma\mathbf{T}$ is a triple (ξ, η, H) where $(\xi, \eta) : f \to g$ is a morphism in $\mathbf{E}(2,4)$, see (3.2.1), and where H is a homomorphism

$$\begin{cases} H : B_2' \to \tilde{\otimes}^4 A_2 \text{ with} \\ \delta H h = -\bar{\Gamma}(\eta)\bar{f} + \bar{g}\xi. \end{cases} \tag{2}$$

For this we consider the following diagram where Hh is given by the quotient map $h : E_{B'} \to B' \to B_2'$.

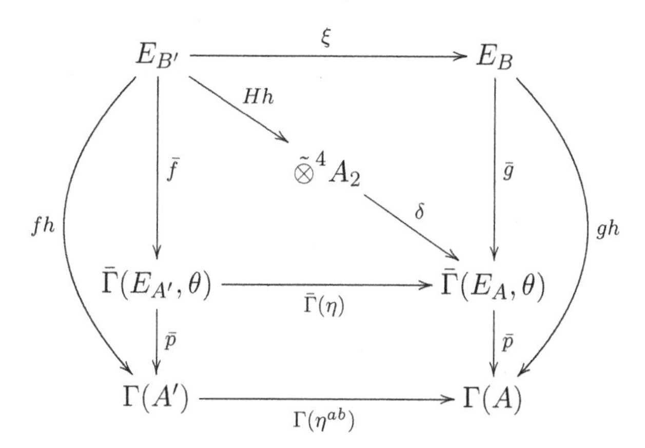

Since $g(\xi^{ab}) = \Gamma(\eta^{ab})f$ by (3.2.1), we see that $\bar{p}(-\bar{\Gamma}(\eta)\bar{f} + \bar{g}\xi) = 0$. Hence by exactness of (5.1.18) there exists H satisfying (2). This shows that the forgetful functor p which carries (ξ, η, H) to (ξ, η) is full. We define the composition in $\Gamma\mathbf{T}$ for $(\xi', \eta', H') : \bar{f}' \to \bar{f}$ with $f' : B'' \to \bar{\Gamma}(A'')$ by

$$(3) \qquad \begin{cases} (\xi, \eta, H)(\xi', \eta', H') = (\xi\xi', \eta\eta', H * H'), \\ H * H' = \tilde{\otimes}^4(\eta_2)H' + H\xi_2' + \Theta(\eta, \eta')(\tau f')_2. \end{cases}$$

Here $\Theta(\eta, \eta')$ and τ are defined as in (5.1.16); by (5.1.17) we see that the composition in (3) is well defined. One can check that the composition law (3) is associative so that $\Gamma\mathbf{T}$ is a well defined category. Finally we obtain the action of $\alpha \in E_\Gamma(\xi, \eta) = Hom(B_2', \Gamma(A) \otimes \mathbb{Z}/2)$ by

$$(4) \qquad (\xi, \eta, H) + \alpha = (\xi, \eta, H + i\alpha)$$

where i is the inclusion in (5.1.18). Using the exactness of (5.1.18) one readily checks that (4) yields a well defined linear extension of categories as described in (1).

The next result shows that the category $\Gamma\mathbf{T}$ is an algebraic model of the topological track category $\mathbf{T}\Gamma$ in (3.3.1)(2).

5.1.20. THEOREM: *There is an isomorphism*

$$\chi : \mathbf{T}\Gamma \overset{\cong}{\to} \Gamma\mathbf{T}$$

of categories which yields an equivalence of the linear extensions (5.1.19)(1) and (3.3.1)(2).

This result is proved in (7.7) below. Since $\mathbf{T}(2,4)$ is a pull back category as in (3.3.3) we see that (5.1.20) and (4.1.7) yield an algebraic model of the category $\mathbf{T}(2,4)$; we will consider this model in more detail in the next section, (5.2).

Similarly as in (4.1.9) we have the following result on the quotient category $\mathbf{CW}(2,4)/D_L$, see (4.1.8).

5.1.21. THEOREM: *There is an equivalence of categories* $\mathbf{CW}(2,4)/D_\Gamma \xrightarrow{\sim} \Gamma\mathbf{T}/\simeq$ *where the notation* $\simeq$ *on* $\Gamma\mathbf{T}$ *is defined as follows.*

5.1.22. DEFINITION: We set $(\xi,\eta,H) \simeq (\xi',\eta',H')$ if $\xi^{ab} = (\xi')^{ab}$, $\eta^{ab} = (\eta')^{ab}$ and if there exist homotopies H_η, H_ξ with

$$\begin{aligned} &H_\eta \in T_2(\eta,\eta'), \quad H_\eta : A' \to A \otimes A, \\ &H_\xi \in T_3(\xi,\xi'), \quad H_\xi : B' \to B \hat{\otimes} B \end{aligned}$$

as in (3.1.3) such that the map $\alpha : B' \to \Gamma(A) \otimes \mathbb{Z}/2$ which we define below satisfies $q(\alpha) = 0$ in $D(\xi,\eta)/D_L(\xi,\eta)$. We obtain α by the formula

$$(1) \qquad -\bar{H}_\eta f + H + g_\# H_\xi = H' + i\alpha.$$

Here i is the inclusion in (5.1.18) and $\bar{H}_\eta$ is defined in (5.1.18)(3). Moreover we set

$$(2) \qquad g_\# H_\xi = I(\tau g \hat{\otimes} \tau g) H_\xi$$

where $\tau g : B \to \Gamma A \to \otimes^2 A$ and where I is given by

$$\begin{cases} I : (\otimes^2 A)\hat{\otimes}(\otimes^2 A) \to \tilde{\otimes}^4 A_2, \\ I(x \otimes y)\hat{\otimes}(x' \otimes y') = xyx'y'. \end{cases}$$

By use of (5.1.18)(4) amd (5.1.11)(1) on can see that the equation

$$(3) \qquad \delta(-\bar{H}_\eta f + H + g_\# H_\xi) = \delta H'$$

is satisfied so that α in (1) is well defined.

5.2. Algebraic models of the categories $\mathbf{T}(2,4)$ and $\mathbf{CW}(2,4)$

In this section we describe one of the main results of this book. Using the quadratic refinement $\bar{\Gamma}$ of Whitehead's functor Γ we obtain an algebraic model $\mathbf{T}$ of the topological track category $\mathbf{T}(2,4)$. Using this category $\mathbf{T}$ we also get an algebraic model of the category $\mathbf{CW}(2,4)$. The results in this section are proved in chapter 7 by use of the theory of quadratic chain complexes in [BCH]. We define an algebraic track category $\mathbf{T}$ by the pull back diagram of categories

(5.2.1)
$$\begin{array}{ccc} \mathbf{T} & \xrightarrow{q} & \Gamma\mathbf{T} \\ {\scriptstyle q}\downarrow & \mathit{pull} & \downarrow{\scriptstyle p} \\ \mathbf{LT} & \xrightarrow[p]{} & \mathbf{E}(2,4) \end{array}$$

where $\Gamma\mathbf{T}$ and $\mathbf{LT}$ are defined in (5.1.19) and (4.1.6) respectively. Here $\mathbf{LT} \to \mathbf{E}(2,4)$ is a split extension so that $\mathbf{T} \to \Gamma\mathbf{T}$ is also a split linear extension. Therefore we define $\mathbf{T}$ in terms of $\Gamma\mathbf{T}$ as follows.

5.2.2. DEFINITION: Let $\mathbf{T}$ be the following category. Objects g in $\mathbf{T}$ are the same as in $\mathbf{H}(2,4)$ or $\mathbf{E}(2,4)$, that is g is a homomorphism $g : B \to \Gamma(A)$ where B and A are free abelian groups. A morphism $f \to g$ in $\mathbf{T}$ with $f : B' \to \Gamma(A')$ is a quadruple (ξ, η, H, α) where $(\xi, \eta) : f \to g$ is a morphism in $\mathbf{E}(2,4)$ and $(\xi, \eta, H) : f \to g$ is a morphism in $\Gamma\mathbf{T}$ and $\alpha \in Hom(B', L(A,1)_3)$. Composition is defined by

$$(\xi, \eta, H, \alpha)(\xi', \eta', H', \alpha') = (\xi\xi', \eta\eta', H * H', \alpha * \alpha').$$

Here $H * H'$ is defined as in (5.1.19)(3) and

$$\alpha * \alpha' = \alpha(\xi')^{ab} + L(\eta^{ab}, 1)_3\alpha'.$$

The next result shows that the category $\mathbf{T}$ is an algebraic model of the topological track category $\mathbf{T}(2,4)$ in (3.3.3).

5.2.3. THEOREM: *There is an isomorphism*

$$\chi : \mathbf{T}(2,4) \overset{\cong}{\to} \mathbf{T}$$

of categories which yields an equivalence of linear extensions of $\mathbf{E}(2,4)$.

This is a consequence of (5.1.20) and (4.2.6). We obtain by theorem (5.2.3) the next result which yields an algebraic model of the category $\mathbf{CW}(2,4)$. For this we introduce in (5.2.5) below the homotopy relation $\simeq$ on the category **T** which via the homomorphism (5.2.3) corresponds to the homotopy relation $\simeq$ on $\mathbf{T}(2,4)$ in (3.2.3). Theorem (5.1.21) is a consequence of the following result.

5.2.4. THEOREM: *There is an equivalence of categories*

$$\chi : \mathbf{CW}(2,4) \xrightarrow{\sim} \mathbf{T}/\simeq$$

where the notion of homotopy $\simeq$ *on* **T** *is defined as follows.*

5.2.5. DEFINITION: We set $(\xi,\eta,H,\alpha) \simeq (\xi',\eta',H',\alpha')$ if $\xi^{ab} = (\xi')^{ab}$, $\eta^{ab} = (\eta')^{ab}$ and if there are homotopies H_η, H_ξ with

(5.2.6) $$H_\eta \in T_2(\eta,\eta'), H_\eta : A' \to A \otimes A,$$

(2) $$H_\xi \in T_3(\xi,\xi'),\ H_\xi : B' \to B\hat{\otimes}B$$

as in (3.1.3), such that the map

(3) $$\alpha = (\alpha_L,\alpha_\Gamma) : B' \to L(A,1)_3 \oplus \Gamma(A)\otimes\mathbb{Z}/2 = \Gamma_2^2(A)$$

which we define below satisfies $q(\alpha) = 0$ in $D(\xi,\eta)$. Here we use the quotient map $q : E(\xi,\eta) \to D(\xi,\eta)$ which we derive from (1.3.3), see (3.2.1)(6). We obtain α by the formulas

(4) $$-\tilde{H}_\eta f + H(\xi,\eta) = H(\xi',\eta') + i(-\alpha+\alpha',\alpha_L).$$

Here $H(\xi,\eta)$ and $H(\xi',\eta')$ are defined by the section s as in (4.5.3). Formula (4) determines the coordinate α_L of α. Moreover the coordinate α_Γ of α is given by the formula

(5) $$-\bar{H}_\eta f + G + g_\# H_\xi = G' + i\alpha_\Gamma.$$

Here $\tilde{H}_\eta, \bar{H}_\eta$ and $g_\# H_\xi$ are defined in (4.1.5)(3), (5.1.18)(3) and (5.1.22)(2) respectively. We point out that H_η in (1) appears in both equations (4) and (5).

5.2.7. REMARK: Assume $\eta = \eta'$ in (5.2.5) so that H_η is a homomorphism of the form

$$(1) \qquad H_\eta : A' \xrightarrow{\alpha} \Gamma A \overset{T}{\hookrightarrow} A \otimes A.$$

In this case we get

$$(2) \qquad \tilde{H}_\eta f + \bar{H}_\eta f : B' \xrightarrow{\beta} \Gamma_2^2 A \overset{i}{\hookrightarrow} \otimes_2^2 A$$

with

$$(3) \qquad -\beta = \bar{q}(\alpha \otimes \mathbb{Z}/2)\sigma f + \bar{q}(\alpha \otimes \eta^{ab})\tau f.$$

Here $\bar{q}$ is the quotient map given by (1.1.9). The formula in (3) corresponds to the formula for $d(\alpha)$ in (1.3.3). One can check (3) by the definition of β in (2):

Proof (of (3)): We may assume that $f : B' = \Gamma A$ is the identity. Let $x' \in A'$ and let $H_\eta(x') = x \otimes x$ so that $\alpha(x') = \gamma(x)$. Moreover let $\eta^{ab}x' = y$. Then we get

$$\begin{aligned}
\tilde{H}_\eta(\gamma x') &= (\eta^{ab} \otimes H_\eta - H_\eta \otimes \eta^{ab})\tau x' \\
&= y \otimes x \otimes x - x \otimes x \otimes y \\
&= [[y, x], x]
\end{aligned}$$

$$\begin{aligned}
\bar{H}_\eta(\gamma x') &= [m(H_\eta \otimes \eta^{ab}) + n(\eta^{ab} \otimes H_\eta) + I(H_\eta \otimes H_\eta)](x' \otimes x') \\
&= m(x \otimes x \otimes y) + n(y \otimes x \otimes x) + I(x \otimes x \otimes x \otimes x) \\
&= xxyx + yxyx + xxxx \\
&= i([x, y] \otimes 1 + (\gamma x) \otimes 1).
\end{aligned}$$

Here we use $xxyx = xyxy$, see (5.1.5)(4). On the other hand we get

$$
\begin{aligned}
\beta(\gamma x') &= [\bar{q}(\alpha \otimes \mathbb{Z}/2)\sigma + \bar{q}(\alpha \otimes \eta^{ab})\tau](\gamma x') \\
&= \bar{q}(\alpha(x') \otimes 1) + \bar{q}(\alpha(x') \otimes y) \\
&= (\gamma x) \otimes 1 + [\gamma x, y] \\
&= (\gamma x) \otimes 1 + [x, y] \otimes 1 - [[y, x], x].
\end{aligned}
$$

In the last equation we use the Barcus–Barrat–formula (1.1.9)(iii). In fact, we can consider the formulas (5.2.5)(3),(4) as the analogue (on the level of tracks) of the Barcus–Barrat–formula (which is given on the level of homotopy classes). □

5.2.8. REMARK: The category $\mathbf{P}(2,4)_0$ is a subcategory of $\mathbf{CW}(2,4)$ and therefore theorem (5.2.4) yields also an algebraic model category equivalent to $\mathbf{P}(2,4)_0$. On the other hand we have computed $\mathbf{P}(2,4)_0$ completely in theorem (2.7.3) above. It remains a challenging problem to derive the result in (2.7.3) from theorem (5.2.4). This would give a proof of (2.7.3) independent of Rochlin's result [R] used implicitly in the proof of (2.7.3).

CHAPTER 6

CROSSED CHAIN COMPLEXES AND ALGEBRAIC MODELS OF TRACKS

The crossed chain complex $\rho(X)$ of a CW-complex X was introduced by J.H.C Whitehead [W]. In chapter III of the book [BCH] one finds the basic theory of crossed chain complexes which is used in the proofs of this chapter. In particular we shall use the crossed chain complex $\rho(JX)$ of the James construction JX where X is a one point union of 1-spheres. The 3-type of $\rho(JX)$ leads to the definition of the universal G-quadratic pair in (6.1.1), see (4.1.1). This 3-type carries all the information which is needed for the computation of the quotient category $\mathbf{CW}(2,4)/D_\Gamma$ in (4.5.6). In chapter 7 we shall study the "quadratic chain complex" $\sigma(JX)$ which is then used for the computation of $\mathbf{CW}(2,4)$.

6.1. The quadratic refinement $\tilde{\Gamma}$

In (4.1.1) we introduced the refinement $\tilde{\Gamma}$ of Whitehead's functor Γ. In this section we prove some of the properties already described in (4.1) and we show that $\tilde{\Gamma}(G)$ is a G^{ab}-module. For the convenience of the reader we recall the definition of $\tilde{\Gamma}(G)$ as follows.

6.1.1. Definition: Let G be a group (not necessarily abelian) and let A be an abelian group. We call a function f and a homomorphism F as in the diagram

$$G \xrightarrow{f} A \xleftarrow{F} G^a ab \otimes G^{ab} \otimes G^{ab}$$

a *G-quadratic pair* if the properties (1)... (4) belowhold. We set

$$\begin{aligned} [x,y,z] &= F(\{x\} \otimes \{y\} \otimes \{z\}), \\ [x,y] &= f(x+y) - f(x) - f(y) \end{aligned}$$

for $x, y, z \in G$.

$$(1) \qquad [x, x, y] = [y, x, x],$$

$$(2) \qquad f(-x) = f(x) + 2[x, x, x],$$

$$(3) \qquad [x + y, z] = [x, z] + [y, z] - \Delta(x, y, z),$$

$$(4) \qquad [x, y + z] = [x, y] + [x, z] - \Delta(x, y, z),$$

where $\Delta(x, y, z) = [x, z, y] + 2[z, y, x] + 3[z, x, y]$.

If $F = 0$ then f is just a quadratic map in the sense of (1.1.1). A G-quadratic pair

$$(5) \qquad G \xrightarrow{\tilde{\gamma}} A \xleftarrow{\delta} G^{ab} \otimes G^{ab} \otimes G^{ab}$$

is *universal* if for each G-quadratic pair (f, F) as above there is a unique homomorphism

$$(6) \qquad (f, F)^{\square} : \tilde{\Gamma}(G) \to A$$

with $(f, F)^{\square}\tilde{\gamma} = f$ and $(f, F)^{\square}\delta = F$. The universal G-quadratic pair (5) exists and yields a functor

$$(7) \qquad \tilde{\Gamma} : \mathbf{Gr} \to \mathbf{Ab}.$$

For a homomorphism $\alpha : G' \to G$ the induced homomorphism $\tilde{\Gamma}(\alpha)$ is given by

$$(8) \qquad \tilde{\Gamma}(\alpha) = (\tilde{\gamma}\alpha, \delta(\alpha^{ab} \otimes \alpha^{ab} \otimes \alpha^{ab}))^{\square}$$

where we use (6).

We now consider some properties of the universal quadratic pair $(\tilde{\Gamma}(G), \tilde{\gamma}, \delta)$ for G. We first observe that $\tilde{\Gamma}$ actually depends only on the quotient $G/\Gamma_3 G$ of nilpotency degree 2.

6.1.2. PROPOSITION: *Let $\Gamma_3 G$ be the subgroup in G generated by all triple commutators in G. Then the quotient map $q : G \to G/\Gamma_3 G$ induces an isomorphism*

$$q_* : \tilde{\Gamma}(G) \xrightarrow{\cong} \tilde{\Gamma}(G/\Gamma_3 G).$$

Proof: It is enough to show that each G-quadratic pair (f, F) admits a factorization

$$f : G \overset{q}{\twoheadrightarrow} G/\Gamma_3 G \xrightarrow{f'} A. \tag{1}$$

Then (f', F) is a G-quadratic pair and the diagram

(2)

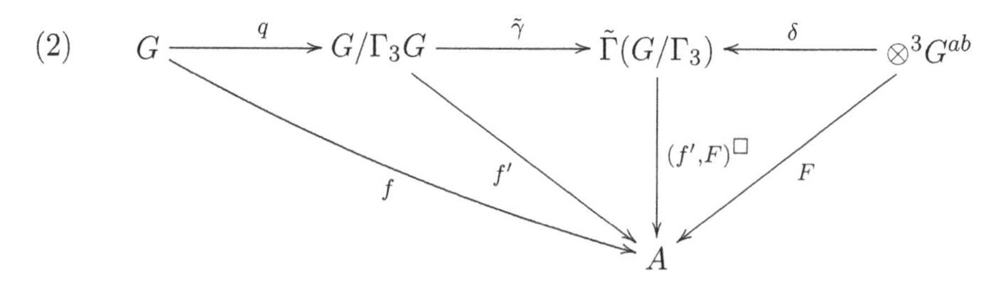

shows that $(\tilde{\gamma}q, \delta)$ is the universal G quadratic pair. The factorization (1) is obtained by the following lemma. □

6.1.3. LEMMA: *Let (f, F) be a G-quadratic pair. Then for $x, y, z \in G$ the formula*

$$f(z + (x, y)) = f(z) + [x, y, z] - [z, x, y]$$

holds where $(x, y) = -x - y + x + y$ is the commutator in G.

Proof: We first observe that (f, F) has the following properties.

$$0 = f(0) = [0, x] = [x, 0]. \tag{1}$$

Moreover

$$[x, x] = 2f(x) - 4[x, x, x]. \tag{2}$$

In fact, we get (2) as follows. We have

$$\begin{aligned} 0 &= f(-x+x) = f(-x) + f(x) + [-x,x] = 2f(x) + 2[x,x,x] + [-x,x], \\ 0 &= [-x+x,x] = [-x,x] + [x,x] - [-x,x,x] - 2[x,x,-x] - 3[x,-x,x] \end{aligned}$$

so that

$$[-x,x] = -[x,x] - 6[x,x,x]. \tag{3}$$

This implies (2). More generally we get

$$[-x,y] = -[x,y] - [x,y,x] - 5[y,x,x], \tag{4}$$

$$[x,-y] = -[x,y] - 3[x,y,y] - 3[y,x,y]. \tag{5}$$

We now show that the lemma holds for $z = 0$, that is $f(-x-y+x+y) = 0$.

$$\begin{aligned} f(-(y+x)+(x+y)) &= f(-(y+x)) + f(x+y) + [-(y+x), x+y] \\ &= f(x+y) + 2[y+x,y+x,y+x] + f(x+y) \\ &\quad - [y+x,x+y] - [y+x,x+y,y+x] \\ &\quad - 5[x+y,y+x,y+x] \\ &= f(y+x) + f(x+y) \\ &\quad - [y+x,x+y] \qquad (6) \\ &\quad - 4[y+x,y+x,y+x] \qquad (7) \\ &= f(y) + f(x) + f(x) + f(y) + [y,x] + [x,y] \\ &\quad + (6) + (7) \\ &= [x,x] + 4[x,x,x] + [y,y] + 4[y,y,y] + [y,x] + [x,y] \qquad (8) \\ &\quad - [y,x+y] - [x,x+y] + (7) \\ &\quad + [y,x+y,x] + 2[x+y,x,y] + 3[x+y,y,x] \qquad (9) \\ &= (7) + (8) + (9) - [x,y] - [y,y] \\ &\quad + [y,y,x] + 2[y,x,y] + 3[y,y,x] \qquad (10) \\ &\quad - [x,x] - [x,y] \\ &\quad + [x,y,x] + 2[y,x,x] + 3[y,x,x] \qquad (11) \\ &= 4[x,x,x] + 4[y,y,y] + (7) + (9) + (10) + (11) \\ &= 0. \end{aligned}$$

In the last equation we use multilinearity and the rule (1) in (6.1.1). Next we obtain

$$f(z + (x,y)) = f(z) + f((x,y)) + [z,(x,y)] = f(z) + [z,(x,y)].$$

Here we have

$$\begin{array}{rcl} [z,(x,y)] & = & [z,-(y+x)+(x+y)] \\ & = & [z,-(y+x)]+[z,x+y] \\ (12) & & +\ 3[z,x+y,x+y]+3[x+y,z,x+y] \\ & = & -\ [z,y+x]+[z,x+y] \\ & & -\ 3[z,y+x,y+x]-3[y+x,z,y+x]+(12) \\ & = & -\ [z,y+x]+[z,x+y] \\ (13) & = & \Delta(z,y,x)-\Delta(z,x,y) \\ & = & F(A). \end{array}$$

By definition of Δ we get

$$A = zxy + 2xyz + 3xzy - zyx - 2yxz - 3yzx$$

where we set $xyz = x \otimes y \otimes z$. Now one can check, see (1.1.8)(2), that

$$(14) \qquad A - xyz - zxy \ = \ [[x,y],z] + 3[[z,x],y]$$

where $F[[x,y,z]=0$ by the following lemma. Whence the proof of the formula is complete. □

6.1.4. LEMMA: *Let (f,F) be a G-quadratic pair. Then F vanishes on triple Lie brackets in $\otimes^3 G^{ab}$, that is $F[[x,y],z]=0$.*

Proof: By (1) in (6.1.1) we know

$$\begin{array}{rcl} [x+y,x+y,z] & = & [z,x+y,x+y], \\ {[x,x,z]=[z,x,x]} & \text{and} & [y,y,z]=[z,y,y] \end{array}$$

so that

$$[x,y,z]+[y,x,z]=[z,x,y]+[z,y,x].$$

This is equivalent to $F[[x,y],z]=0$, compare the definition of triple Lie brackets in (1.1.8)(2). □

6.1.5. PROPOSITION: *The abelian group $\tilde{\Gamma}(G)$ is in a canonical way a G^{ab}-module. For $\alpha \in G^{ab}$ we define the action of α on $a \in \tilde{\Gamma}(G)$ by $a^\alpha =$*

$(\alpha_\#, \delta)^\square(a)$ *where* $\alpha_\# : G \to \tilde{\Gamma}(G)$ *is given by*

$$\alpha_\#(x) = \tilde{\gamma}(x) - [x, x, \alpha] = \tilde{\gamma}(x)^\alpha.$$

Proof: We first check that $(\alpha_\#, \delta)$ ia a G-quadratic pair so that $(\alpha_\#, \delta)^\square$ is defined. In fact we have

$$\begin{aligned} \alpha_\#(-x) &= \tilde{\gamma}(-x) - [-x, -x, \alpha] \\ &= \tilde{\gamma}(x) + 2[x, x, x] - [x, x, \alpha] \\ &= \alpha_\#(x) + 2[x, x, x]. \end{aligned}$$

Moreover we get

$$\begin{aligned} [x, y]_\alpha &= -\alpha_\#(y) + \alpha_\#(x + y) - \alpha_\#(x) \\ &= [x, y] - [x + y, x + y, \alpha] + [x, x, \alpha] + [y, y, \alpha] \\ &= [x, y] - [x, y, \alpha] - [y, x, \alpha] \end{aligned}$$

where $[x, y, \alpha] - [y, x, \alpha]$ is linear in x and y. This completes the proof that $(\alpha_\#, \delta)$ is a G-quadratic pair. It is clear that $a^0 = a$. Moreover $a^{\beta+\alpha} = (a^\alpha)^\beta$ follows from $(\beta_\#, \delta)^\square \alpha_\# = (\alpha + \beta)_\#$. In fact

$$\begin{aligned} (\beta_\#, \delta)^\square \alpha_\#(x) &= (\beta_\#, \delta)^\square(\tilde{\gamma}(x) - [x, x, \alpha]) \\ &= \beta_\#(x) - [x, x, \alpha] \\ &= \tilde{\gamma}(x) - [x, x, \beta] - [x, x, \alpha] \\ &= (\alpha + \beta)_\#(x). \end{aligned}$$

□

6.2. The crossed James construction of a group

The crossed James construction $J(G)$ of a group G is motivated by the crossed chain complex $\rho J(X)$ of the classical James construction of the space X. If X is a one point union of 1-spheres with $\pi_1 X = G$ we have the isomorphism $J(G) \cong \rho J(X)$ which is a crucial ingredient in the proof of the main results of chapter 4.

We first introduce the following notation. A *G-group* M is given by an action of the group G on the group M denoted by x^α for $x \in M$, $\alpha \in G$. We have

$$\begin{aligned}
(x+y)^\alpha &= x^\alpha + y^\alpha, \\
(-x)^\alpha &= -x^\alpha, \\
x^{\alpha+\beta} &= (x^\alpha)^\beta, \\
x^0 &= x.
\end{aligned}$$

A G-group M is a *G-module* if M is abelian. For a homomorphism $\varphi : G \to G'$ in **Gr** a φ-equivariant map $F : M \to M'$ from a G-group M to a G'-group M' is a homomorphism between groups satisfying $F(x^\alpha) = F(x)^{\varphi(\alpha)}$. Let $\mathbf{Mod}_{\mathbb{Z}}^{\wedge}$ be the following category: objects are pairs (G, M) where M is a G-module and morphisms $(\varphi, F) : (G, M) \to (G', M')$ are φ-equivariant maps F.

Recall that each CW-complex X with $X^0 = *$ yields a crossed chain complex $\rho(X)$ given by the boundary maps

$$\text{(6.2.1)} \qquad \ldots \xrightarrow{d} \pi_3(X^3, X^2) \xrightarrow{d} \pi_2(X^2, X^1) \xrightarrow{d} \pi_1(X^1)$$

of relative homotopy groups. This is the "homotopy system" of X considered by J.H.C. Whitehead in [W]. The crossed chain complex $\rho(X)$ is totally free in the sense that it satisfies a freeness condition in each degree, the basis of $\rho(X)$ is given by the set of cells of $X - *$. In particular $\rho(X)_1 = \pi_1(X^1)$ is a free group generated by the 1-cells of X. Let **cross chain** be the category of crossed chain complexes (which need not be totally free). This category is studied in [BCH]. We intoduce a functor

$$\text{(6.2.2)} \qquad J : \mathbf{Gr} \to \mathbf{cross\ chain}$$

which we call the *crossed James construction* on the category of groups. For a group G the crossed chain complex $J(G)$ is generated by all words $a_1 \ldots a_n$ ($a_i \in G$, $i = 1, \ldots, n$ and $n \geq 1$) with the defining relations (1), (2) and (3) below plus, of course, the laws of crossed chain complexes; let u, v be such words or empty and let $a, a', t \in G$; then (a) denotes the word given by the single letter a; moreover the degree $\mid u \mid$ of the word $u = a_1 \ldots a_n$ is n, that is $u \in J(G)_n$:

$$\text{(1)} \qquad (a)^t = (a^t) \text{ where } a^t = -t + a + t.$$

(2)
$$u(a+a')v = \begin{cases} (a)+(a') & \text{for } \; u=\emptyset=v, \\ ua'v+(uav)^{a'} & \text{for } \; |u|\geq 1, \\ (uav)^{a'}+ua'v & \text{for } \; |v|\geq 1. \end{cases}$$

(3)
$$d(a)=0 \text{ and } d(uv) = \begin{cases} -u-v+u+v & \text{for } \; |u|=|v|=1, \\ -v^u+v-u(dv) & \text{for } \; |u|=1, \; |v|\geq 2, \\ (du)v+(-1)^{|u|}(-u^v+u) & \text{for } \; |u|\geq 2, \; |v|=1, \\ (du)v+(-1)^{|u|}u(dv) & \text{for } \; |u|\geq 2, \; |v|\geq 2. \end{cases}$$

For a homomorphism $F: G' \to G$ the induced homomorphism $J(F): J(G') \to J(G)$ is defined by $J(F)(a_1 \dots a_n) = (Fa_1)\dots(Fa_n)$.

Now let X be a CW-complex with $X^0 = *$. The *James consruction* $JX = \varinjlim J_nX$ is obtained by the quotient space $J_nX = X^n/\sim$ where X^n is the n-fold product with the CW-topology and where the equivalence relation $\sim$ is generated by $(x_1,\dots,x_{n-1}) \sim (x_1,\dots,x_{t-1},*,x_t,\dots,x_{n-1})$ with $t=1,\dots,n$ and $x_t \in X$. The space JX is a CW-complex with cells $e_1 \times \dots \times e_n$ where e_i is a cell in $X - *$, $i=1,\dots,n$ and $n \geq 1$. It is a classical result of James [J] that there is a natural homotopy equivalence

(6.2.3)
$$J(X) \xrightarrow{\simeq} \Omega\Sigma X$$

where $\Omega\Sigma X$ denotes the loop space of the suspension of X. We now consider the case that

(6.2.4)
$$X = K(G,1) = M_A = \bigvee_Z S^1$$

is a one point union of 1-spheres S^1, here $G = \pi_1 X =< Z >$ is the free group generated by the set Z and $A = G^{ab}$ is the abelianization of G, see the notation in (3.1.2). The crossed James construction on the group G in (6.2.2) is justified by the following result.

6.2.5. THEOREM: *For a free group $G =< Z >$ there is a natural isomorphims of crossed chain complexes*

$$\psi : \rho JK(G,1) \cong J(G).$$

Here the left hand side is the crossed chain complex of the CW-complex $JK(G,1)$. The isomorphism ψ carries the product cell $e_1 \times \dots \times e_n$ to the word $e_1 \cdots e_n$ where we identify a 1–cell e in $K(G,1) = \bigvee_Z S^1$ with the corresponding generator $e \in Z$ in G. The theorem is proved in III.C.6 of [BCH]. Let

$$\pi_n(\rho) = kernel(d : \rho_n \to \rho_{n-1})/image(d : \rho_{n+1} \to \rho_n)$$

be the homology of the crossed chain complex $\rho = (\cdots \to \rho_2 \to \rho_1)$. Then we have for a CW-complex X the well known isomorphism $\pi_1 \rho X = \pi_1 X$ and $\pi_n \rho X = H_n \hat{X}$, $n \geq 2$, where $\hat{X}$ is the universal covering of X. We use this isomorphism for the proof of the next result.

6.2.6. PROPOSITION: *Let G be a free group with $A = G^{ab}$. Then one has natural isomorphisms*

$$\begin{aligned} \pi_1 J(G) &= A, \\ \pi_2 J(G) &= \Gamma(A), \\ \pi_3 J(G) &= L(A,1)_3. \end{aligned}$$

Proof: We clearly have for $X = M_A$ in (6.2.4)

$$\text{(1)} \qquad \pi_2 J(G) = H_2 \widehat{JX} = \pi_2 JX = \pi_2 \Omega \Sigma X = \pi_3 \Sigma X = \Gamma(A)$$

by (1.1.6). Here $\widehat{JX}$ is the universal covering of JX. Moreover Whitehead's Γ-sequence yields the exact row

$$\text{(2)} \qquad \begin{array}{ccccccc} \Gamma\pi_2(JX) & \xrightarrow{i} & \pi_3(JX) & \xrightarrow{\hat{h}} & H_3(\widehat{JX}) & \longrightarrow & 0 \\ \| & & \| & & \| & & \\ \Gamma\Gamma(A) & \xrightarrow{\sigma} & \Gamma_2^2(A) & \twoheadrightarrow & L(A,1)_3 & & \end{array}$$

Here the left hand isomorphism is given by (1) and the isomorphism in the middle is obtained by $\pi_3(JX) = \pi_4 \Sigma X$, compare (1.1.9). Since the operator i is induced by the Hopf map $\eta : S^3 \to S^2$ the left hand part of the diagram (2) commutes, compare the property of σ in (1.1.7). This implies that one gets the isomorphism on the right hand side of (3), so that

$$\text{(3)} \qquad \pi_3 J(G) = H_3 \widehat{JX} = L(A,1)_3.$$

$\square$

For any group G we clearly have $(JG)_1 = G$. Moreover there is a commutative diagram of G-groups

(6.2.7)
$$\begin{array}{ccccccc}
\cdots \longrightarrow & (JG)_3 & \xrightarrow{d} & (JG)_2 & \xrightarrow{d} & G \\
 & \downarrow q & & \downarrow q & & \downarrow q \\
\cdots \xrightarrow{0} & \otimes^3(G^{ab}) & \xrightarrow{0} & \otimes^2 G^{ab} & \xrightarrow{0} & G^{ab}
\end{array}$$

where q is the surjection which carries the word $a_1 \dots a_n$ to $\{a_1\} \otimes \cdots \otimes \{a_n\}$. The boundary maps in the bottom row are the trivial maps 0 and G acts trivially on $\otimes^n G^{ab}$.

We now construct the following natural commutative diagram:

(6.2.8)
$$\begin{array}{ccccccc}
 & & G & & & & \\
 & & \downarrow f & & & & \\
(JG)_3 & \xrightarrow{d} & \hat{J}G & \rightarrowtail & (JG)_2 & \xrightarrow{d} & G \\
h_3 = q \downarrow & (i) & \downarrow & (ii) & \downarrow h_2 & & \| \\
\otimes^3 G^{ab} & \xrightarrow{\delta} & \tilde{J}G & \rightarrowtail & (\tilde{J}G)_2 & \xrightarrow{\tilde{d}} & G
\end{array}$$

Here $\hat{J}G = kernel(d : (JG)_2 \to G)$ and (i) is a push out diagram, (ii) is a central push out diagram (see [BCH]). The induced crossed module $\tilde{d}$ satisfies $\tilde{J}G = kernel(\tilde{d})$. The function f is defined by

$$f(a) = aa.$$

We clearly have $df = 0$ by (6.2.2)(3). The next result on the map δ in (6.2.8) is a corollary of (6.2.6).

6.2.9. COROLLARY: *For a free group G one has the natural exact sequence*

$$L(G^{ab}, 1)_3 \overset{i}{\hookrightarrow} \otimes^3 G^{ab} \xrightarrow{\delta} \tilde{J}(G) \overset{\tilde{p}}{\twoheadrightarrow} \Gamma(G^{ab}).$$

Proof: We consider the commutative diagram

$$\begin{array}{ccccccc}
L(A,1)_3 & \rightarrowtail & J(G)_3/im(d) & \xrightarrow{d} & \hat{J}G & \twoheadrightarrow & \Gamma(A) \\
\| & & \downarrow q & (i) & \downarrow & & \| \\
L(A,1) & \underset{q'}{\rightarrowtail} & \otimes^3 A & \underset{\delta}{\longrightarrow} & \tilde{J}G & \twoheadrightarrow & \Gamma(A)
\end{array}$$

where $A = G^{ab}$ and where (i) is a push out diagram by (6.2.8). The map q' can be identified with

$$q' = p_* : H_3\widehat{JX} \to H_3JX$$

where p is the projection of the universal covering. Moreover $q'\hat{h}$, with $\hat{h}$ in (6.2.6)(2) above, can be identified with the James–Hopf invariant γ_3 in (1.1.9)(5). Therefore q' is injective. Since the top row of the diagram above is exact, by (6.2.6), the bottom row is also exact, since (i) is a push out by definition of $\tilde{J}G$. □

6.3. The isomorphism $\tilde{J}G = \tilde{\Gamma}G$ for a free group G

In this section we consider the pair of maps

$$\otimes^3 G^{ab} \xrightarrow{\delta} \tilde{J}G \xleftarrow{h_2 f} G$$

defined by (6.2.8) and we show that this is a quadratic pair in the sense of (6.1.1). This in fact is the original motivation for the definition of quadratic pairs. Moreover we show

6.3.1. THEOREM: *For a free group G the quadratic pair $(\delta, h_2 f)$ above is universal so that $\tilde{J}G \cong \tilde{\Gamma}G$.*

As a corollary we obtain the main part of theorem (4.1.4). For the proof of (6.3.1) we need the following auxiliary notation; the proof of (6.3.1) is given in (6.3.10) below.

6.3.2. DEFINITION: Let G be a group and let A be a G^{ab}-module. We say that a map $f : G \to A$ is *strongly quadratic* if the properties (1) and (2) below hold where we define $\mid x, y \mid$ by the equation

$$\mid x, y \mid = f(x+y)^{-y} - f(x)^y - f(y)^{-y}$$

for $x, y \in G$. Here the action of G on A is given by the action of G^{ab}, that is $a^x = a^{\{x\}}$ where $\{x\} \in G^{ab}$ is the class of x and $a \in A$.

$$f(-x) = f(x)^{-2x}. \tag{1}$$

The function $| \_, \_ |: G \times G \to A$ is a bicrossed homomorphism, that is

$$\begin{cases} | x+y, y | &= | x, z |^y + | y, z |, \\ | x, y+z | &= | x, y |^z + | x, z |. \end{cases} \tag{2}$$

If A is a trivial G^{ab}-module (that is, if $a^x = a$) then a strongly quadratic map is the same as a quadratic map in the sense of (1.1.1). A strongly quadratic map

$$\tilde{\gamma} : G \longrightarrow \hat{\Gamma}(G) \tag{3}$$

is *universal* if for each strongly quadratic map f as above there is a unique homomorphism of G^{ab}-modules

$$f^{\square} : \hat{\Gamma}(G) \longrightarrow A \tag{4}$$

with $f^{\square}\hat{\gamma} = f$. The universal strongly quadratic map (3) yields a functor

$$\hat{\Gamma} : \mathbf{Gr} \to \mathbf{Mod}_{\mathbb{Z}}^{\wedge}. \tag{5}$$

For a homomorphism $\alpha : G' \to G$ the induced homomorphism $\hat{\Gamma}(\alpha)$ is given by $\hat{\Gamma}(\alpha) = (\hat{\gamma}\alpha)^{\square}$. Clearly $\hat{\Gamma}(\alpha)$ is an α^{ab}-equivariant homomorphism.

6.3.3. LEMMA: *A strongly quadratic map $f : G \to A$ satisfies*

$$f(0) = 0 = | 0, x | = | x, 0 | \textit{ and } | x, x | = 2f(x).$$

Proof: $0 = f(-x + x) = | -x, x |^x + f(-x)^{2x} + f(x) = -| x, x | + 2f(x)$. □

In the next two propositions we describe two examples of strongly quadratic maps.

6.3.4. PROPOSITION: *The function $f : G \to \hat{J}G$ in (6.2.8) is strongly quadratic.*

Proof: We have by (6.2.2) the following equations with $x, y, z \in G$.

$$(1) \qquad \begin{aligned} f(x+y) &= (x+y)(x+y) \\ &= (x(x+y))^y + y(x+y) \\ &= (xy + (xx)^y)^y + yy + (yx)^y \\ &= f(x)^{2y} + f(y) + \mid x, y \mid^y \end{aligned}$$

where we set

$$(2) \qquad \mid x, y \mid = xy + yx.$$

Moreover we get

$$(3) \qquad \begin{aligned} \mid x+y, z \mid &= (x+y)z + z(x+y) \\ &= (xz)^y + yz + zy + (zx)^y \\ &= \mid x, z \mid^y + \mid y, z \mid, \text{ and} \end{aligned}$$

$$(4) \qquad \begin{aligned} \mid x, y+z \mid &= x(y+z) + (y+z)x \\ &= xz + (xy)^z + (yx)^z + zx \\ &= \mid x, y \mid^z + \mid x, z \mid. \end{aligned}$$

Finally we observe by (1)...(4) that $f(0) = 0$ and $\mid 0, x \mid = \mid x, 0 \mid = 0$. Moreover

$$(5) \qquad \begin{aligned} 0 &= f(x-x) = f(x)^{-2x} + f(-x) + \mid x, x \mid^{-x}, \\ &= f(x)^{-2x} + f(-x) - \mid x, x \mid^{-2x} \ \text{(see(4))} \\ &= f(x)^{-2x} + f(-x) - (2f(x))^{-2x} \ \text{(see(2))} \\ &= -f(x)^{-2x} + f(-x). \end{aligned}$$

This completes the proof that f is strongly quadratic. □

6.3.5. PROPOSITION: *The map $\tilde{\gamma} : G \to \tilde{\Gamma}(G)$ of the universal G-quadratic pair is a strongly quadratic map. Here we use the structure of $\tilde{\Gamma}(G)$ as a G^{ab}-module in (6.1.5).*

Proof: We have

$$(1) \qquad \tilde{\gamma}(-x) = \tilde{\gamma}(x) - [x, x, -2x] = \tilde{\gamma}(x)^{-2x}$$

by (6.1.1)(2) and (6.1.5). Moreover we get

$$
(2)\quad \begin{aligned} | x, y | &= \tilde{\gamma}(x+y) + [x+y, x+y, y] \\ &\quad -(\tilde{\gamma}(x) - [x, x, y]) - (\tilde{\gamma}(y) + [y, y, y]) \\ &= [x, y] + 2[x, x, y] + [x, y, y] + [y, x, y]. \end{aligned}
$$

For $z \in G$ we deduce from (6.1.5) and the definition of $| x, y |$ the formula

$$
(3)\quad | x, y |^z = | x, y | - [x, y, z] - [y, x, z].
$$

Now (2), (3) and (6.1.1)(3) show that $| \_, \_ |$ is a bicrossed homomorphism. For this we use (6.1.4). □

The next result is a partial inverse of (6.3.5).

6.3.6. PROPOSITION: *Let $f : G \to A$ be a strongly quadratic map and suppose a homomorphism $F : \otimes^3 G^{ab} \to A$ is given satisfying*

(a) $[x, x, y] = [y, x, x]$,
(b) $(fx)^y = f(x) - [x, x, y]$

for $x, y \in G$ with $[x, y, z] = F(\{x\} \otimes \{y\} \otimes \{z\})$. Then (f, F) is a G-quadratic pair.

Proof: The conditions (1) and (2) in (6.1.1) for (f, F) are obviously satified. It remains to check (3) and (4) in (6.1.1). We have by (6.3.2) and (b) the equation

$$
\begin{aligned} | x, y |^z &= -f(y)^{-y+z} + f(x+y)^{-y+z} - f(x)^{y+z} \\ &= -f(y) + [y, y, -y+z] + f(x+y) - [x+y, x+y, -y+z] \\ &\quad -f(x) + [x, x, y+z] \end{aligned}
$$

so that

$$
(1)\quad | x, y |^z = | x, y | - [x, y, z] - [y, x, z].
$$

Hence by definition of $| x, y |$ and (b) we have the equation

$$
f(x+y) = | x, y |^y + f(x)^{2y} + f(y) = | x, y | - [x, y, y] - [y, x, y] + f(x) - [x, x, 2y] + f(
$$

so that

$$(2) \qquad [x,y] = | x,y | - 2[x,x,y] - [x,y,y] - [y,x,y]$$

as in (6.3.5)(2). Now (2) and (6.3.2)(2) show that $[x+y,z]$ satisfies the formula in (6.1.1)(3). For this we also need (a) and the argument in the proof of (6.1.4). □

6.3.7. COROLLARY: *The pair $(h_2 f, \delta)$ in (6.3.1) is a quadratic pair.*

Proof: Since $h_2 : \hat{J}G \to \tilde{J}G$ is G^{ab}-equivariant $h_2 f$ is also strongly quadratic. Hence we can use (6.3.6) for the proof that $(h_2 f, F)$ is a G-quadratic pair. For this we consider the boundary in JG:

$$(1) \qquad \begin{aligned} d(xxz) &= d(xx)z + (-1)^{|xx|}(-(xx)^z + xx) \\ &= -f(x)^z + f(x). \end{aligned}$$

This implies for $[x,y,z] = \delta h_3(xyz)$, see (6.2.8),

$$(2) \qquad f(x)^z = f(x) - [x,x,z].$$

Finally

$$(3) \qquad d(xxy) = -f(x)^y + f(x) = d(yxx), \text{ see (6.2.2)(3)},$$

implies $[x,x,y] = [y,x,x]$. □

Proposition (6.3.5) implies that one has a natural commutative diagram

$$(6.3.8) \qquad \begin{array}{ccccc} \hat{\Gamma}(G) & \xrightarrow{p} & \tilde{\Gamma}(G) & \xrightarrow{\tilde{p}} & \Gamma(G^{ab}) \\ & \nwarrow^{\hat{\gamma}} & \uparrow^{\tilde{\gamma}} & & \uparrow^{\gamma} \\ & & G & \xrightarrow[q]{} & G^{ab} \end{array}$$

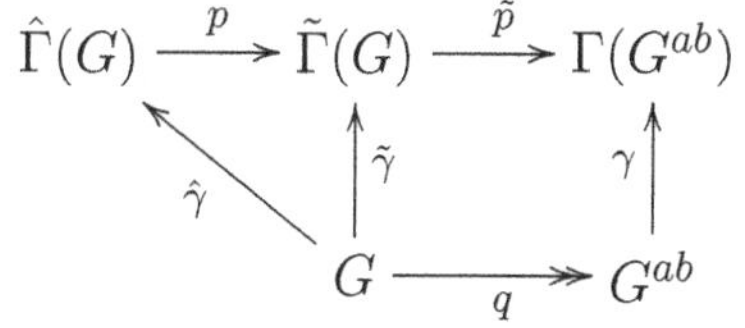

where $p = (\tilde{\gamma})^{\square}$ and $\hat{p} = \tilde{p}p = (\gamma q)^{\square}$, compare (6.3.2). The maps $p, \tilde{p}$ and $\hat{p}$ are equivariant with respect to the action of G^{ab}, here $\Gamma(G^{ab})$ has the trivial action of G^{ab}. We shall use the following property of the composition $\hat{p} = \tilde{p}p$.

6.3.9. PROPOSITION: *Suppose G is a free group, then the elements $\hat{\gamma}x - (\hat{\gamma}x)^{\alpha}$, $x \in G$, $\alpha \in G^{ab}$, generate the kernel of $\hat{p} : \hat{\Gamma}(G) \to \Gamma(G^{ab})$ as a G^{ab}-module.*

Proof. As a G^{ab}-module $\hat{\Gamma}(G)$ is generated by the elements $\hat{\gamma}(g)$, $g \in G$. Let Z be a basis of the free group $G = < Z >$ and choose a well ordering $<$ of Z. Then the elements

$$\gamma(e),\ e \in Z,\ \text{and } [e, e'],\ e < e',\ e, e' \in Z, \tag{1}$$

form a basis of the free $\mathbb{Z}$-module $\Gamma(G^{ab})$. Let $s : \Gamma(G^{ab}) \to \hat{\Gamma}(G)$ be the homomorphism defined by $s\gamma(e) = \hat{\gamma}(e)$ and $s[e, e'] = \mid e, e' \mid$. Then s is a section of $\hat{p}$. Now let $R \subset \hat{\Gamma}(G)$ be the subgroup generated by all elements $a - a^{\alpha}$, $a \in \hat{\Gamma}(G)$, $\alpha \in G^{ab}$. Then (6.3.2)(1),(2) show that

$$\hat{\gamma}(g) - s\gamma(qg) \in R \text{ for } g \in G. \tag{2}$$

Here $q : G \to G^{ab}$ is the projection. We see this by induction on the length of the reduced word g in $< Z >= G$. In fact, if (2) is true for g then (2) also holds if we replace g by $g+e$ or $g-e$, $e \in Z$. For $\epsilon \in \{+1, -1\}$ we get modulo R

$$\begin{aligned} & \hat{\gamma}(g + \epsilon e) - s\gamma q(g + \epsilon e) \\ = \ & \mid g, \epsilon e \mid^{\epsilon e} + \hat{\gamma}(g)^{\epsilon e} + \hat{\gamma}(\epsilon e) - s\gamma q(g) - s\hat{\gamma} e - \epsilon s[g, e] \\ \equiv \ & \mid g, \epsilon e \mid - \epsilon s[g, e]. \end{aligned}$$

Again we see by induction on the length of g that the term (3) is an element in R. Hence the proof of (2) is complete. Moreover (2) shows that

$$R = kernel(\hat{p} : \hat{\Gamma}(G) \to \Gamma(G)). \tag{4}$$

Since $a \in \hat{\Gamma}(G)$ is a sum $a = \sum_i (\hat{\gamma} x_i)^{\alpha_i}$ with $x_i \in G$, $\alpha_i \in G^{ab}$ we get

$$a - a^{\alpha} = \sum_i ((\hat{\gamma} x_i)^{\alpha_i} - (\hat{\gamma} x_i)^{\alpha_i + \alpha}) = \sum_i (\hat{\gamma} x_i - \hat{\gamma} x_i^{\alpha})^{\alpha_i}. \tag{5}$$

This shows that R is generated as a G^{ab}-module by the element $\hat{\gamma} x - (\hat{\gamma} x)^{\alpha}$, $x \in G$, $\alpha \in G^{ab}$. □

6.3.10. Proof (of theorem (6.3.1)): We obtain by (6.3.4) the commutative diagram

(1)
$$\begin{array}{ccc} \hat{\Gamma}(G) & \xrightarrow{\ p\ } & \tilde{\Gamma}(G) \\ {\scriptstyle f^{\square}}\downarrow & & \downarrow{\scriptstyle (h_2 f,\delta)^{\square}=k} \\ \hat{J}(G) & \xrightarrow[\ h_2\ ]{} & \tilde{J}(G) \end{array}$$

where $p = \tilde{\gamma}^{\square}$ as in (6.3.8). The composition $kp = h_2 f^{\square}$ of maps in (1) is used in the following commutative diagram where we construct an inverse k' of k in (1).

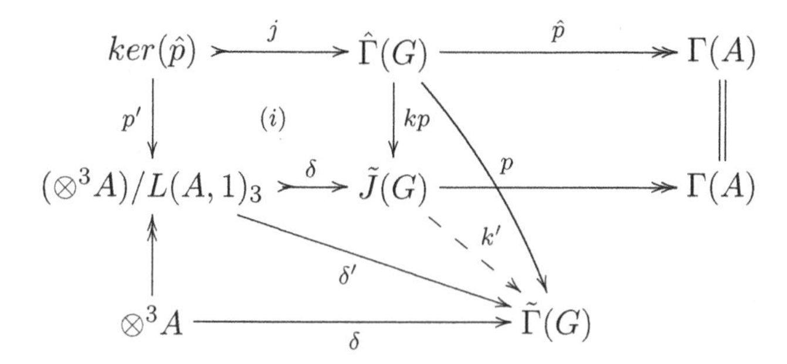

Here the rows are short exact sequences so that (i) is a push out diagram. The map ∂ is induced by δ in (6.2.8). We now show that

(2)
$$pj = \delta' p'$$

where δ' is given by δ in (6.1.1) via (6.1.4) and where p' is the restriction of kp. For the proof of equation (2) we use (6.3.9) so that it is enough to check

$$\begin{aligned} & p(\hat{\gamma}x - (\hat{\gamma}x)^{\alpha}) && = \tilde{\gamma}(x) - (\tilde{\gamma}x)^{\alpha} \\ = \ & \tilde{\gamma}x - \tilde{\gamma}x + [x,x,\alpha] && = [x,x,\alpha]. \end{aligned}$$

$$\begin{aligned} \delta' p'(\hat{\gamma}x - (\hat{\gamma}x)^{\alpha}) &= \delta' \partial^{-1} kp(\hat{\gamma}x - (\hat{\gamma}x)^{\alpha}) \\ &= \delta' \partial^{-1} k[x,x,\alpha] \\ &= \delta' \partial^{-1} \partial(\{x\} \otimes \{x\} \otimes \alpha) \\ &= \delta'(\{x\} \otimes \{x\} \otimes \alpha\} = [x,x,\alpha]. \end{aligned}$$

Whence (6.3.9) shows that (2) holds. Therefore there is a unique homomorphism k' which extends the diagram above commutatively. Since $k'\delta = \delta'$ and

$k'kp\hat{\gamma} = k'h_2f$ we see that k' is the inverse of k in (1). Moreover k is an isomorphism since k is surjective; this follows from the exact sequence (6.2.9). □

6.4. The crossed 3-type of a loop space

We apply the crossed chain complex functor ρ to sets of topological tracks. This yields algebraic models of tracks which we use for the computation of the track category $\mathbf{TL} = \mathbf{T}(2,4)/E_\Gamma$, compare theorem (4.1.7).

Let ρ be a totally free crossed chain complex. We say that a crossed chain complex ρ' is the *n-type* of ρ if $(\rho')_j = 0$ for $j > n$ and if there is a map $h : \rho \to \rho'$ which induces isomorphisms $\pi_j h : \pi_j \rho \cong \pi_j \rho'$ for $j < n$. We now consider the 2-type and the 3-type of the crossed chain complex $J(G)$ where G is a free group. Here $J(G)$ is the crossed chain complex of the loop space $\Omega M(A,2) \simeq J(M_A)$ with $A = G^{ab}$. Therefore we call the 3-type of J(G) the *crossed* 3-*type of the loop space* $\Omega M(A,2)$.

6.4.1. THEOREM: *Let G be a free group with $A = G^{ab}$, $E_A = G/\Gamma_3 G$. Then the commutator map (see (3.1.3))*

$$wq : \otimes^2 A \to E_A \tag{1}$$

is the "2-type of J(G)". Moreover the bottom row in (6.2.8)

$$\otimes^3 A \xrightarrow{\delta} (\tilde{J}G)_2 \xrightarrow{\tilde{d}} G \tag{2}$$

is the "3-type of J(G)". By (6.3.1) we have the natural isomorphism kernel$(\tilde{d})$ $\cong \tilde{\Gamma}(G)$ which we use as an identification. Thus δ in (2) is given by δ in (6.1.1)(5).

Proof: The 2-equivalence h from $J(G)$ to wq is given by $q : (JG)_2 \to \otimes^2 A$ in (6.2.7) and by the quotient map $(JG)_1 = G \to E_A$. The 3-equivalence h from $J(G)$ to the crossed chain complex $(\delta, \tilde{d})$ in (2) is given by the maps in diagram (6.2.8), see (6.2.9). □

Proposition (3.1.4) is a corollary of (6.4.1)(1) above since we can define the bijection χ for $n = 2$ as follows. Let $x, y : M_{A'} = X \to M_A = Y$ be maps between one point unions of 1-spheres with $G' = \pi_1 X$, $G = \pi_1 Y$. Then we get

$$\chi : T(\Sigma x, \Sigma y) = [I_* \Sigma X, \Sigma Y]^{\Sigma x, \Sigma y} \tag{6.4.2}$$

$$\begin{array}{lll}
(1) & \cong & [I_*X, \Omega\Sigma Y]^{ix,iy} \\
(2) & \cong & [I_*X, JY]^{jx,jy} \\
(3) & \cong & [\rho I_*X, \rho JY]^{x_*,y_*} \\
(4) & \cong & [I_*G', JG]^{x_*,y_*} \\
(5) & \cong & [I_*G', wq]^{hx_*,hy_*} \\
(6 & \cong & T_2(\xi,\eta).
\end{array}$$

The map $i : Y \to \Omega\Sigma Y$ in (1) is adjoint to the identity of ΣY and $j : Y \hookrightarrow JY$ is the inclusion. We obtain the bijection (2) by the homotopy equivalence (6.2.3). Moreover bijection (3) is induced by the functor ρ in (6.2.1), for this we use III.7.2 in [BCH]. The maps $x_* = \rho(x) = \pi_1(x)$ and $y_* = \rho(y) = \pi_1(y)$ are induced by the functor ρ. The cylinder I_*G' in (4) is the cylinder in the category of crossed chain complexes where G' is concentrated in degree 1, see III.3.3 in [BCH]. Moreover we use (6.2.5) in (4). The bijection (5) is induced by the 2-equivalence h in the proof of (6.4.1) and (6) is the obvious bijection since ξ (resp. η) is induced by x_* (resp. y_*).

Similarly as in (6.4.2) we obtain the equivalence χ of categories in (4.1.7). For this we may assume that the object $g : M(B,3) \to M(A,2)$ is the adjoint of a composition

$$M(B,2) \xrightarrow{\bar{g}} J(M_A) \xrightarrow{\cong} \Omega M(A,2) \tag{6.4.3}$$

where $\bar{g}$ is a cellular map. We now consider the set of tracks H in (3.2.1)(3). We get the following composition of maps where $X = \Sigma(t\xi)$ and $y = t\eta$, $X = M(B',2)$, $Y = M_A$.

$$\begin{array}{lll}
\chi : T(g(\Sigma x), (\Sigma y)f) & = & [I_*\Sigma X, \Sigma Y]^{g(\Sigma x),(\Sigma y)f} \\
(1) & \cong & [I_*X, JX]^{\bar{g}x, J(y)\bar{f}} \\
(2) & \twoheadrightarrow & [\rho I_*X, \rho JY]^{(\bar{g}x)_*,(J(y)\bar{f})_*} \\
(3) & \cong & [I_*B', JG]^{\bar{g}_*x_*, J(y_*)\bar{f}_*} \\
(4) & \cong & [I_*B', (\delta,\tilde{d})]^{h\bar{g}_*x_*, hJ(y_*)\bar{f}_*} \\
(5) & \cong & [I_*B', \delta]^{\tilde{g}\xi^{ab}, \tilde{\Gamma}(\eta)\tilde{f}}.
\end{array}$$

Here the bijection (1) is given by the homotopy equivalence (6.2.3) similarly as in (6.4.2)(1),(2). The surjection (2) is defined by the functor ρ in (6.2.1) and $(\bar{g}x)_* = \rho(\bar{g}x)$ is the induced map. Now the cylinder I_*B' is the cylinder of the chain complex B' which is concentrated in degree 2. We clearly have $I_*B' = \rho I_*X$. Therefore the bijection (3) is given by the isomorphism (6.2.5). The bijection (4) is induced by the 3-equivalence h for (6.4.1)(2). This immediately shows that one gets the bijection (5) where $\tilde{g}$ is the unique map for which the diagram

(6)
$$\begin{array}{ccc} B = (\rho M(B,2))_2 & \xrightarrow{\bar{g}_*} & (\rho J M_A)_2 = (JG)_2 \\ \tilde{g}\downarrow & & \downarrow h_2 \\ \tilde{\Gamma}(E_A) \cong \tilde{\Gamma}(G) \cong ker(\tilde{d}) & \rightarrowtail & (\tilde{J}G)_2 \end{array}$$

commutes, see (6.2.4). The set in (5) denotes a set of tracks in the category of chain complexes, in particular, $\delta : \otimes^3 A \to \tilde{\Gamma}(E_A)$ is considered as a chain complex concentrated in degrees 2 and 3. It is clear that the elements of the set (5) are exactly all homomorphisms $H : B' \to \otimes^3 A$ for which $(\xi, \eta, H) : \tilde{f} \to \tilde{g}$ is a morphism in the category **LT**, see (4.1.6)(2). We now define the isomorphism of categories

(6.4.4)
$$\chi : \mathbf{TL} = \mathbf{T}(2,4)/E_\Gamma \xrightarrow{\cong} \mathbf{LT}$$

by $\chi(\xi, \eta, H) = (\xi, \eta, \chi(H))$ where we use the map χ in (1)...(5) above. It is clear that $\chi(H + \alpha) = \chi(H)$ for $\alpha \in E_\Gamma(\xi, \eta) = Hom(B', \Gamma(A) \otimes \mathbb{Z}/2)$ so that χ in (6.4.4) is well defined. Moreover χ in (6.4.4) is a bijection on morphism sets by (6.2.6)(2) where $\hat{h}$ is also induced by the functor ρ.

We now consider the composition

(6.4.5)
$$I_*M(B',3) \xrightarrow{I_*f} I_*M(A',2) \xrightarrow{H_\eta} M(A,2)$$

where $H_\eta : \Sigma y \simeq \Sigma y'$, $y_* = \eta$, $y'_* = \eta'$. By (6.4.2) the homotopy H_η is equivalently given by a homomorphism $H_\eta : \otimes^2 A' \to A$ with $H_\eta \in T_2(\eta, \eta')$. The adjoint of H_η is a map

(1)
$$\bar{H}_\eta : I_*M_{A'} \longrightarrow J(M_A)$$

and the adjoint of $H_\eta(I_*f)$ in (6.4.5) is given by the composition

(2)

$$\tilde{H}_\eta : I_*M(B',2) \xrightarrow{I_*\bar{f}} I_*JM_{A'} \xrightarrow{K} J(I_*M_{A'}) \xrightarrow{J\bar{H}_\eta} JJM_A \xrightarrow{r} JM_A.$$

Here $K : J(i_1) \simeq J(i_2)$ is a homotopy for which $i_1, i_2 : M_{A'} \hookrightarrow I_*M_{A'}$ are the inclusions. Moreover r is the canonical retraction defined by multiplication of words in JM_A. We may assume that K is a cellular map. We want to compute the induced map $\alpha = \rho(K)$ which is a homotopy

(3) $$\alpha : I_*JG' = \rho I_*JM_{A'} \xrightarrow{\rho K} \rho J(I_*M_{A'}) = J(I_*G'),$$

with $G' = \pi_1 M_{A'}$. Here G' is a crossed chain complex concentrated in degree 1 and I_*G' is the cylinder. For $G' =< Z' >$ we obtain I_*G' by the free crossed module

(4) $$d_2 : (I_*G')_2 \longrightarrow (I_*G')_1 = G * G$$

where $G * G$ is the free product of groups with inclusions $i_1, i_2 : G \to G * G$. We write $i_1 x = x'$, $i_2 x = x''$, for $x \in G$. The basis of the source of d_2 is the suspension sZ' with

(5) $$d_2(se) = -e' + e'' \text{ for } e \in Z'.$$

We clearly have $I_*G' = \rho(I_*M_{A'})$, compare III.3.3 in [BCH]. In the right hand side of (3) we use the James construction of a crossed chain complex and theorem III.C.6 [BCH]. The map α in (3) corresponds to a homotopy

(6) $$\alpha : J(i_1) \simeq J(i_2),\ \alpha_i : (JG')_i \longrightarrow (JI_*G')_{i+1},$$

where $i_1, i_2 : G' \to I_*G'$ are the inclusions, compare III.3.3 in [BCH] and III.2.6(3) in [BCH]. Here α_1 is the i_1-crossed homomorphism with $\alpha_1(e) = se$ for $e \in Z'$. Moreover we define α_2 to be the i_1-equivariant homomorphism satisfying

(7) $$\alpha_2(xy) = (\alpha_1 x)(i_1 y) - (i_2 x)(\alpha_1 y)$$

for $x, y \in G$, compare (6.2.2). The multiplication on the right hand side of (7) is defined by the multiplication in $J(I_*G'')$, compare III.C.3 in [BCH]. One can check that α_2 in (7) is well defined, for this replace x by $x + x'$ and use the rules (6.2.2)(2) and III.C.4(3) in [BCH].

(8)**Lemma:** There is a homotopy α as in (6) with α_2 satisfying (7). Moreover K in (2) can be chosen such that (3) holds for α.

Proof: The map ρK in (3) and a homotopy α as in (6) are both maps between cylinders; such maps are unique up to homomorphism relative to the boundary of the cylinder. This follows from the lifting lemma II.1.11 in [BAH], see also II.2.10 in [BAH]. Therefore it is enough to prove that α in (7) can be extended to a homotopy α as in (6). This, in fact, follows by showing that in degree 3 the equation

$$-i_1 + i_2 = d\alpha_2 + \alpha_1 d \tag{9}$$

holds, see III.2.6 in [BCH]. We set $i_1 x = x'$ and $i_2 x = x''$ and $\alpha_1 x = \underline{x}$. For $x, y \in G'$ and $xy \in (JG')_2$ we thus have

$$(-i_1 + i_2)(xy) = -x'y' + x''y''. \tag{10}$$

On the other hand we get

$$\begin{aligned} \alpha_1 d(xy) &= \alpha_1(-x - y + x + y) \text{ (see (6.2.2)(3))} \\ &= -\underline{x}^{d(x'y')} - \underline{y}^{x'+d(x'y')} + \underline{x}^{y'} + \underline{y} \\ &= -x'y' - \underline{x} - \underline{y}^{x'} + x'y' + \underline{x}^{y'} + \underline{y} \end{aligned} \tag{11}$$

since α_1 is an i_1-crossed homomorphism and since (11) is an equation in a crossed module. Hence we have to find $\alpha_2(xy)$ in such way that (9) or equivalently

$$d\alpha_2(xy) = x''y'' - B - x'y' + \underline{y}^{x'} + \underline{x} \text{ with } B = \underline{x}^{y'} + \underline{y} \tag{12}$$

holds. Using III.C.4 [BCH] we get

$$
\begin{aligned}
d(x'' \cdot \underline{y}) &= -\underline{y}^{x''} + \underline{y} - x''(-y' + y'') \\
&= -\underline{y}^{x''} + \underline{y} - (x''y'' - x''(-y')^{y''}) \\
&= -\underline{y}^{x''} + \underline{y} + (x''y')^{-y'+y''} - x''y'' \\
&= -\underline{y}^{x''} + \underline{y} - \underline{y} + x''y' + \underline{y} - x''y'' \\
&= -\underline{y}^{x''} + x''y' + \underline{y} - x''y''.
\end{aligned}
\tag{13}
$$

Similarly we have

$$
d(\underline{x}y') = -x'y' + \underline{x} + x''y' - \underline{x}^{y'}. \tag{14}
$$

Now (13) and (14) imply for $A = x''\underline{y} - \underline{x}y'$

$$
dA = -\underline{y}^{x''} - \underline{x} + x'y' + B - x''y''. \tag{15}
$$

For this we replace $x''y'$ in (13) by $x''y'$ in (14). We now replace B in (12) by B in (15). This yields the equation

$$
\begin{aligned}
d\alpha_2(xy) &= x''y'' - x''y'' - dA - \underline{y}^{x''} - \underline{x} + x'y' - x'y' + \underline{y}^{x'} + \underline{x} \\
&= -\underline{y}^{x''} - \underline{x} + \underline{y}^{x'} + \underline{x} - dA \\
&= -\underline{y}^{x''} + \underline{y}^{x'+(-x'+x'')} - dA \\
&= -dA
\end{aligned}
$$

which is equivalent to (12). Hence for $\alpha_2(xy) = -A$ equation (12), and thus also (9), is satisfied. This proves formula (7). □

Using (7) and (8) above we get the next result on the induced map $\rho(\tilde{H}_\eta)$ with $\tilde{H}_\eta$ given by (2) above.

6.4.6. PROPOSITION: *Let h be the 3-equivalence in the proof of (6.4.1). Then the map $h\rho(\tilde{H}_\eta)$ satisfies in degree 3 the equation*

$$
(h\rho(\tilde{H}_\eta))_3 = (H_\eta \otimes \eta^{ab} - \eta^{ab} \otimes H_\eta)\tau f : B' \to \otimes^3 A.
$$

Proof: We first observe that α_1 in (6.4.5)(7) yields the composition

(1) $$G' \xrightarrow{\alpha_1} (I_* G')_2 \xrightarrow{\rho \bar{H}_\eta} (JG)_2 \xrightarrow{q} \otimes^2 A$$

which is a homomorphism and which carries a generator $e \in Z'$ to $H\{e\}$, compare (6.4.5)(1) and (6.4.3)(1)...(5). By (6.4.5)(7) and (1) we see that

(2) $$v = h\rho(r(J\bar{H}_\eta)K) = h \circ \rho(r) \circ J(\rho \bar{H}_\eta) \circ \alpha$$

carries $xy \in J(G')_2$ to

(3) $$v(xy) = H_\eta\{x\} \otimes \eta^{ab}\{y\} - \eta^{ab}\{x\} \otimes H_\eta\{y\}.$$

Hence v coincides with the composition

(4) $$v : (JG')_2 \xrightarrow{q} \otimes^2 A' \xrightarrow{H_\eta \otimes \eta^{ab} - \eta^{ab} \otimes H_\eta} \otimes^3 A.$$

Since $q\rho(\bar{f}) = \tau f$ by (6.4.3)(6) we get (6.4.6). □

CHAPTER 7

QUADRATIC CHAIN COMPLEXES AND ALGEBRAIC MODELS OF TRACKS

The subject of this chapter is the "quadratic refinement" of the results and methods in chapter 6. For this we replace "crossed chain complexes" in chapter 6 by "quadratic chain complexes". One finds the theory of quadratic chain complexes in the book [BCH]. This theory is extensively used in the proofs. In particular, we shall use the quadratic chain complex of the James construction JX where X is a one point union of 1-spheres. The 3-type of this quadratic chain complex is an algebraic equivalent of the topological 3-type of the loop space $\Omega M(A,2) \simeq JX$ where $M(A,2) = \Sigma X$ is the Moore space of a free abelian group A. We derive from the algebraic model of this 3-type the algebraic model of the homotopy category $\mathbf{CW}(2,4)$ of (2,4)-complexes described in chapter 5.

7.1. Square groups and the functor $\overset{\approx}{\Gamma}$

We introduce a functor $\overset{\approx}{\Gamma}$ which is defined on the category of θ-square–groups. The definition of θ-square groups is motivated by the properties of the tensor product of quadratic chain complexes below. We show that there are natural transformations $\overset{\approx}{\Gamma} \twoheadrightarrow \tilde{\Gamma}$ and $\overset{\approx}{\Gamma} \twoheadrightarrow \bar{\Gamma}$ where $\tilde{\Gamma}$ and $\bar{\Gamma}$ are the quadratic refinements of Γ in (4.1.5) and (5.1.11). These transformations lead to a proof of the exact sequence for $\bar{\Gamma}$ in (5.1.15) since a similar exact sequence is available for $\overset{\approx}{\Gamma}$.

7.1.1. DEFINITION: A θ-square–group (G,θ) consists of a group G and a function

$$(1) \qquad \theta : G \longrightarrow \otimes^2 G^{ab}$$

which satisfies for $x, y \in G$

(2) $$\theta(x+y) = \theta(x) + \theta(y) + hx \otimes hy$$

where $h : G \to G^{ab}$ is the abelianization with $hx = \{x\}$. We also write $xy = hx \otimes hy$. Let (G', θ') be a further square group and let $\eta : G' \to G$ be a homomorphism. Then one gets the homomorphism

(3) $$\begin{cases} \theta(\eta) & : \quad G' \xrightarrow{h} (G')^{ab} \xrightarrow{\theta(\eta)} \otimes^2 G^{ab} \\ \theta(\eta) & = \quad -(\otimes^2 \eta^{ab})\theta' + \theta\eta. \end{cases}$$

Let **SGr** be the following category. Objects are θ-square groups and morphisms $\eta : (G', \theta') \to (G, \theta)$ are homomorphisms $\eta : G' \to G$. Let $\mathbf{SGr}_0$ be the subcategory of **SGr** consisting of morphisms η with $\theta(\eta) = 0$. We obtain for each square group (G, θ) the symmetric θ-group (G, q, θ) by use of the quotient map

(4) $$q : \otimes^2 G^{ab} \to \hat{\otimes}^2 G_2.$$

This yields the functors $q_{\#} : \mathbf{SGr} \to \mathbf{SG}$ and $q_{\#} : \mathbf{SGr} \to \mathbf{SG}_0$, see (5.1.8).

7.1.2. LEMMA: *A θ-square group (G, θ) satisfies*

(1) $$\theta(-x - y + x + y) = xy - yx.$$

Therefore θ admits a unique factorization

(2) $$\theta : G \xrightarrow{q} G/\Gamma_3 G \xrightarrow{\theta} \otimes^2 G^{ab}$$

where $\Gamma_3 G$ is the subgroup of triple commutators in G and where q is the quotient map. Moreover the map q yields the quotient map

(3) $$q = (q, 1) : (G, \theta) \to (G/\Gamma_3 G, \theta)$$

in $\mathbf{SGr}_0$.

Proof (of(7.1.2)): We have $\theta(0) = 0$ so that $\theta(-x) = -\theta(x) + xx$. Hence we get

$$\begin{aligned}\theta(-x-y+x+y) &= \theta(-x-y)+\theta(x+y)+(-x-y)(x+y)\\ &= \theta(-x)+\theta(-y)+xy+\theta(x)+\theta(y)+xy\\ &\quad -(x+y)(x+y)\\ &= -\theta(x)+xx-\theta(y)+yy+xy+\theta(x)+\theta(y)+xy\\ &\quad -(x+y)(x+y).\end{aligned}$$

□

7.1.3. EXAMPLE: Let $G =< Z >$ be a free group generated by the set Z with $A = G^{ab}$. Then one has the θ-square group (G, θ_0) where θ_0 is uniquely determined by setting $\theta_0(c) = 0$ for $c \in Z$. Compare (7.2.2) below. Let $E_A = G/\Gamma_3 G$ be the free nil(2)-group given by G and let $q : G \to E_A$ be the quotient map. Then (7.1.2) shows that θ_0 in (1) admits a factorization

$$\theta_0 : G \xrightarrow{q} E_A \xrightarrow{\theta_0} A \otimes A, \tag{2}$$

so that (E_A, θ_0) is again a square group. Moreover q yields the morphism

$$q = (q, 1) : (G, \theta_0) \to (E_A, \theta_0) \tag{3}$$

in $\mathbf{SGr}_0$.

7.1.4. DEFINITION: We define the functor

$$\begin{cases} \tilde{\otimes}^3 : \mathbf{SGr} \to \mathbf{Ab} \text{ by} \\ \tilde{\otimes}^3(G,\theta) = (\otimes^3 G^{ab}) \oplus \Phi(G^{ab}) \text{ where} \\ \Phi(A) = (\otimes^2 A)\hat{\otimes}(\otimes^2 A) \otimes \mathbb{Z}/2. \end{cases} \tag{1}$$

For the functor Φ we use $\hat{\otimes}$ in (1.1.2). The induced map $\tilde{\otimes}^3(\eta)$ for $\eta : G' \to G$ in (7.1.1), however, is given by the coordinates

$$\tilde{\otimes}^3(\eta) = (\otimes^3 \eta^{ab}) \oplus \Phi(\eta^{ab}) + \theta_\#(\eta). \tag{3}$$

Here the correction term $\theta_\#(\eta)$ is the homomorphism $(x, y, z \in G')$

$$(4)\quad \begin{cases} \theta_{\#}(\eta) : \otimes^3(G')^{ab} \to \Phi(G^{ab}), \\ \theta_{\#}(\eta)(\{x\} \otimes \{y\} \otimes \{z\}) = T((\otimes^2\eta^{ab})(xy - yx) \otimes \theta(\eta)(z)) \end{cases}$$

where $\theta(\eta)$ is the homomorphism (7.1.1) and where T is the permutation map

$$(5)\quad \begin{cases} T : (\otimes^2 A) \otimes (\otimes^2 A) \to \Phi(A), \\ T(x_1 \otimes x_2 \otimes y_1 \otimes y_2) = x_2 \otimes y_2 \hat{\otimes} x_1 \otimes y_1 = x_1 \otimes y_1 \hat{\otimes} x_2 \otimes y_2 \end{cases}$$

for $x_1, x_2, y_1, y_2 \in A$. We have the obvious natural exact sequence

$$(6)\quad \Phi(G^{ab}) \overset{w}{\rightarrowtail} \tilde{\otimes}^3(G,\theta) \underset{s}{\overset{p}{\rightleftarrows}} \otimes^3 G^{ab}$$

where w and s are the inclusions and where p is the projection. The splitting s of p, however, is not natural, see (3). It is easy to check that $\tilde{\otimes}^3$ is a well defined functor on **SGr**.

We use the functor $\tilde{\otimes}^3$ for the following generalization of (G,θ)-pairs in (5.1.8). For this we first introduce the following elements.

7.1.5. Definition: Let (G,θ) be a θ-square group with $A = G^{ab}$. We obtain for $x, y, x', y' \in G$ the elements

$$(1)\quad xy = (hx) \otimes (hy) \in A \otimes A \text{ and}$$

$$(2)\quad xy \hat{\otimes} x'y' \in \Phi(A) = (\otimes^2 A) \hat{\otimes} (\otimes^2 A) \otimes \mathbb{Z}/2.$$

Moreover we define the following elements in $\Phi(A)$.

$$(3)\quad \nabla(x,y) = T((xy - yx) \otimes \theta x) + xx \hat{\otimes} yx + yx \hat{\otimes} yx,$$

$$(4)\quad \nabla(x,y,z) = xy \hat{\otimes} zy + T(xy \otimes \theta z + xz \otimes \theta y + yz \otimes \theta x),$$

$$(5)\quad \nabla_1(x,y,z) = \nabla(x,y,z) + \nabla'(x,y,z) \hat{\otimes} zx,$$

$$(6)\quad \nabla_2(x,y,z) = \nabla(x,y,z) + \nabla'(x,y,z) \hat{\otimes} xz \text{ with}$$

$$(7)\quad \nabla'(x,y,z) = xy + yx + zy + yz + yy.$$

The elements ∇_1 and ∇_2 in (5), (6) can also be described as follows. (This is needed in the proof of (7.5.7) below.)

(8) $$U(x,y) = T(xy \otimes (\theta y + \theta x)) + yy\hat{\otimes}xy + xx\hat{\otimes}yx,$$

(9) $$U_1(x,y,z) = T(xy \otimes \theta z) + (yz + zy)\hat{\otimes}zx,$$

(10) $$U_2(x,y,z) = T(yz \otimes \theta x) + (xy + yx)\hat{\otimes}xz,$$

(11) $$\nabla_1(x,y,z) = U_1(x,y,z) + U(x+y,z) - U(x,z) - U(y,z),$$

(12) $$\nabla_2(x,y,z) = U_2(x,y,z) + U(x,y+z) - U(x,y) - U(x,z).$$

7.1.6. DEFINITION: Let (G,θ) be a square group and let K be a group. A (G,θ)-*pair* is a pair (f,F) of functions

$$G \xrightarrow{f} K \xleftarrow{F} \tilde{\otimes}^3(G,\theta)$$

with the following properties. The function F is a homomorphism which maps to the center of K. Moreover, for $x,y,z \in G$ let

(1) $$[x,y,z] = Fs(\{x\} \otimes \{y\} \otimes \{z\}),$$

(2) $$[x,y] = -f(y) + f(x+y) - f(x).$$

In (1) we use the splitting s in (7.1.4)(6). Then f and F satisfy the following equations where we use the elements in (7.1.5).

(3) $$[x,x,y] = [y,x,x] + Fw\nabla(x,y),$$

(4) $$f(-x) = f(x) + s[x,x,x] + FwT(xx \otimes \theta x),$$

(5) $$-f(x) - f(y) + f(x) + f(y) = Fw(xx\hat{\otimes}yy),$$

(6) $$[x+y,z] = [x,z] + [y,z] - \Delta(x,y,z) + Fw\nabla_1(x,y,z),$$

(7) $$[x,y+z] = [x,y] + [x,z] - \Delta(x,y,z) + Fw\nabla_2(x,y,z).$$

Here $\Delta(x,y,z)$ is defined by

$$\Delta(x,y,z) = [x,z,y] + 2[z,y,x] + 3[z,x,y].$$

A (G,θ)-pair

(8) $$G \xrightarrow{\gamma} \tilde{\tilde{\Gamma}}(G,\theta) \xleftarrow{\delta} \tilde{\otimes}^3(G,\theta)$$

is *universal* if for each (G,θ)-pair there is a unique homomorphism between groups

(9) $$(f,F)^{\square} : \tilde{\tilde{\Gamma}}(G,\theta) \longrightarrow K$$

with $(f,F)^{\square}\gamma = f$ and $(f,F)^{\square}\delta = F$. The universal (G,θ)-pair exists and yields a functor (see (7.1.1))

(10) $$\tilde{\tilde{\Gamma}}: \mathbf{SGr}_0 \to \mathbf{Gr}.$$

For a homomorphism $\eta : (G',\theta') \to (G,\theta)$ with $\theta(\eta) = 0$ the induced homomorphism $\tilde{\tilde{\Gamma}}(\eta)$ is given by $\tilde{\tilde{\Gamma}}(\eta) = (\gamma\eta, \delta\tilde{\otimes}^3(\eta))^{\square}$. The group $\tilde{\tilde{\Gamma}}(G,\theta)$ is in a canonical way a G^{ab}-group. For $\alpha \in G^{ab}$ we define the action of α on $a \in \tilde{\tilde{\Gamma}}(G,\theta)$ by $a^{\alpha} = (\alpha_{\#},\theta)^{\square}(a)$ where $\alpha_{\#} : G \to \tilde{\tilde{\Gamma}}(G,\theta)$ is given by

(11) $$\alpha_{\#}(x) = \tilde{\tilde{\gamma}}(x) - [x,x,\alpha] = \tilde{\tilde{\gamma}}(x)^{\alpha}.$$

As in (6.1.5) one shows that $(\alpha_{\#},\delta)^{\square}$ is defined. Moreover, the same argument as in (6.1.5) shows that the G^{ab}-action is well defined.

7.1.7. Definition: Let (G',θ') and (G,θ) be square groups and let $\eta : G' \to G$ be a homomorphism, (that is η is a morphism in $\mathbf{SGr}$). Then we define the function

(1) $$\begin{cases} \nabla_{\eta} : G' \to \phi(G^{ab}), \\ \nabla_{\eta}(x) = T((\otimes^2\eta^{ab})\theta(x) \otimes \theta(\eta)(x)), \ x \in G', \end{cases}$$

see (7.1.1). Consider the following diagram in which the rows are universal quadratic pairs.

(2)
$$\begin{array}{ccccc} G' & \longrightarrow & \widetilde{\widetilde{\Gamma}}(G',\theta') & \longleftarrow & \tilde{\otimes}^3(G',\theta') \\ \downarrow \eta & \overset{f}{\searrow} & \downarrow & \overset{F}{\swarrow} & \downarrow \tilde{\otimes}^3(\eta) \\ G & \longrightarrow & \widetilde{\widetilde{\Gamma}}(G,\theta) & \longleftarrow & \tilde{\otimes}^3(G,\theta) \end{array}$$

(3)
$$\begin{cases} F &= \delta\tilde{\otimes}^3(\eta) \\ f &= \gamma\eta + \delta w \nabla_\eta. \end{cases}$$

One can check that (f,F) is actually a (G',θ')-quadratic–pair so that there is a unique homomorphism

(4)
$$\widetilde{\widetilde{\Gamma}}(\eta) = (f,F)^{\square}.$$

The right hand side of (2) commutes, that is

(5)
$$\widetilde{\widetilde{\Gamma}}(\eta)\delta' = F = \delta\tilde{\otimes}^3(\eta)$$

The left hand side of (2), however, does not commute; here we have

(6)
$$\widetilde{\widetilde{\Gamma}}(\eta)\gamma' = f = \gamma\eta + \delta w \nabla_\eta.$$

If $\theta(\eta) = 0$ the map $\widetilde{\widetilde{\Gamma}}(\eta)$ in (4) coincides with the induced map in (7.1.6). We call $\widetilde{\widetilde{\Gamma}}(\eta)$ the *generalized induced map* for $\widetilde{\widetilde{\Gamma}}$. These generalized induced maps satisfy a composition law as in (7.1.11) below so that they do not define a functor on **SGr**.

We now compare the functor $\widetilde{\widetilde{\Gamma}}$ with the corresponding functors $\tilde{\Gamma}$ and $\bar{\Gamma}$ in (4.1.5) and (5.1.11) respectively. First we consider the functor $\tilde{\Gamma}$. Let (G,θ) be a θ-group and consider the commutative diagram

(7.1.8)
$$\begin{array}{ccccc} G & \xrightarrow{\gamma} & \widetilde{\widetilde{\Gamma}}(G,\theta) & \xleftarrow{\delta} & \tilde{\otimes}^3(G,\theta) \\ \| & & \downarrow h & & \downarrow p \\ G & \xrightarrow{\tilde{\gamma}} & \tilde{\Gamma}(G) & \xleftarrow{\tilde{\delta}} & \otimes^3 G^{ab} \end{array}$$

where the bottom row is the universal G-quadratic–pair in (6.1.1) and where p is the projection in (7.1.4)(6). Then $(\tilde{\gamma}, \tilde{\delta}p)$ is a (G,θ)-pair so that $h = (\tilde{\gamma}, p\tilde{\delta})^{\square}$ is defined. Moreover h is natural with respect to generalized induced maps in (7.1.7), that is $h\tilde{\tilde{\Gamma}}(\eta) = \tilde{\Gamma}(\eta)h$. As in (7.1.10) below one can check that the right hand square of diagram (7.1.8) is a central push out.

Next we consider the functor $\bar{\Gamma}$ in (5.1.11) for which we get the commutative diagram

(7.1.9)
$$\begin{array}{ccccc} G & \xrightarrow{\gamma} & \tilde{\tilde{\Gamma}}(G,\theta) & \xleftarrow{\delta} & \tilde{\otimes}^3(G,\theta) \\ \| & & \downarrow h & & \downarrow Ip \\ G & \xrightarrow{\bar{\gamma}} & \bar{\Gamma}(G,q\theta) & \xleftarrow{\bar{\delta}} & \tilde{\otimes}^4 G_2 \end{array}$$

where the bottom row is the universal $(G, q\theta)$-pair in (5.1.11). Here $(G, q\theta)$ is given by (G, θ) as in (7.1.1)(4). Moreover p in (7.1.9) is the projection $\tilde{\otimes}^3(G,\theta) \to \phi(G^{ab})$ and I is the quotient map in (5.1.4). In (7.1.10) we check that $(\bar{\gamma}, \bar{\delta}Ip)$ is actually a (G,θ)-pair in the sense of (7.1.6) so that the map $h = (\bar{\gamma}, \bar{\delta}Ip)^{\square}$ is defined.

7.1.10. LEMMA: *The map h in (7.1.9) is well defined and natural with respect to generalized induced maps. Moreover the right hand square of diagram (7.1.9) is a central push out and the following diagram commutes.*

(1)
$$\begin{array}{ccc} \tilde{\tilde{\Gamma}}(G,\theta) & \xrightarrow{h} & \tilde{\Gamma}(G) \\ h\downarrow & & \downarrow \tilde{p} \\ \bar{\Gamma}(G,q\theta) & \xrightarrow{\bar{p}} & \Gamma(G^{ab}) \end{array}$$

Here we set $\tilde{\tilde{p}} = \tilde{p}h = \bar{p}h$ where we use $\bar{p}$ and $\tilde{p}$ in (5.1.11) and (4.1.5) respectively. The diagram shows that one has the surjective map

(2)
$$h : \tilde{\tilde{\Gamma}}(G,\theta) \twoheadrightarrow \bar{\tilde{\Gamma}}(G,\theta)$$

given by (h,h) in (1). Here $\bar{\tilde{\Gamma}}$ is a pull back of $(\tilde{p},\bar{p})$. Clearly h in (2) is again natural with respect to generalized induced maps.

Proof: We show that the equations in (5.1.11)(1)...(4) are derived from the corresponding equations (7.1.6)(3)...(7) by projecting the correction terms via Ip to $\tilde{\otimes}^4 G_2$. For this we first observe that

$$IT((xy - yx) \otimes x'y') = 0 \tag{3}$$

and

$$IT(x'y' \otimes (xy - yx)) = 0 \tag{4}$$

holds by (5.1.2)(4),(5), compare (7.1.4)(5). Whence we get by (5.1.5)(4)

$$\begin{aligned} Ip\nabla(x,y) &= xyxy + yxyx \\ &= (xy + yx)(xy + yx) \\ &= i[x,y] \otimes 1 \end{aligned} \tag{5}$$

so that $\bar{\delta} Ip\nabla(x,y) = 0$. Moreover we get

$$Ip\nabla_1(x,y,z) = \nabla'(x,y,z) \tag{6}$$

and

$$Ip\nabla_2(x,y,z) = \nabla''(x,y,z). \tag{7}$$

Here we use for ∇_1 the equation $xyzy = yxzx$ and for ∇_2 the equation $xyzy = zyxy = yzxz$ which follow froms from the relations in (5.1.5). Moreover we use $xyzx = xxzy = zyxx$, $zyzy = xyxz$, $yzzx = zzyx$ and $yyzx = zyyx$ for ∇_1 and $yxxz = xxyz$, $zyxz = xyzz$, $yyxz = xyyz$ for ∇_2. We clearly have $Ip\nabla(x,y,z) = 0$. This completes the proof of (6) and (7). Moreover we get

$$Ipt(xx\hat{\otimes}\theta x) = E(x) = 0 \text{ for } x \in G. \tag{8}$$

This is true for $z \in Z$ since then $\theta z = 0$. Moreover if $E(x) = 0$ and $E(y) = 0$ we get $E(x+y) = 0$ by the relations in (5.1.5). This proves (8). Using (5)...(8) we see that $(\bar{\gamma}, \bar{\delta} Ip)$ is a (G,θ)-pair so that h is well defined. Now consider the central push out Q of (δ, Ip) in (7.1.9). Then the induced maps a, b with

$$G \xrightarrow{\gamma} \tilde{\tilde{\Gamma}}(G,\theta) \xrightarrow{a} Q \xleftarrow{b} \tilde{\otimes} G_2 \tag{9}$$

show that $(a\gamma, b)$ is a $(G, q\theta)$-pair in the sense of (5.1.11). In fact, using the universal property of $\tilde{\tilde{\Gamma}}$ we see that $(a\gamma, b)$ is the universal $(G, q\theta)$-pair. This shows that $q = \bar{\Gamma}(G, q\theta)$. Finally we get the commutativity of (1) since $\tilde{p}h = (\gamma', 0)^{\square}$ and $\bar{p}h = (\gamma', 0)^{\square}$ for the quadratic map $\gamma' : G \to G^{ab} \to \Gamma(G^{ab})$. $\square$

We now consider some further properties of the functor $\tilde{\tilde{\Gamma}}$ above.

7.1.11. LEMMA: *The generalized induced maps in (7.1.7) satisfy the following composition law. For*

$$\eta\eta' : (G'', \theta'') \to (G', \theta') \to (G, \theta) \text{ in } \mathbf{SGr}$$

we have

$$\tilde{\tilde{\Gamma}}(\eta)\,\tilde{\tilde{\Gamma}}(\eta') = \tilde{\tilde{\Gamma}}(\eta\eta') + \delta w \Theta(\eta, \eta')\tau\, \tilde{\tilde{p}}$$

where $\Theta(\eta, \eta') : \otimes^2 (G'')^{ab} \to \phi(G^{ab})$ *is defined by*

$$\Theta(\eta, \eta') = T(\otimes^2 \eta^{ab})\theta(\eta') \otimes \theta(\eta)((\eta')^{ab}),$$

compare (7.1.1).

7.1.12. PROPOSITION: *Let (G, θ) be a square group and let $\Gamma_3 G \subset G$ be the subgroup in G generated by triple commutators. Then one has the quotient map $q : G \to G/\Gamma_3 G$ which by (7.1.2) is a map in $\mathbf{SGr}_0$, and for which the induced map*

$$\tilde{\tilde{\Gamma}}(q) : \tilde{\tilde{\Gamma}}(G, \theta) \xrightarrow{\cong} \tilde{\tilde{\Gamma}}(G/\Gamma_3 G, \theta)$$

is an isomorphism.

As in (6.1.2) the proposition is a consequence of the following lemma.

7.1.13. LEMMA: *Let (f, F) be a (G, θ)-pair. Then for $x, y, z \in G$ the following formulas hold: $f(-x - y + x + y) = F(xyxy)$ and*

$$\begin{aligned} f(z - x - y + x + y) &= f(z) + [x, y, z] - [z, x, y] \\ &\quad + F(xyxy + zxzy + xyzy + T((xz - zx) \otimes \theta y)). \end{aligned}$$

The proof of the lemma is a lengthy but straightforward application of the distributivity laws in (7.1.6), compare (6.1.3). The next result is the analogue of (6.1.5). For this we define the functor Γ_2^2 as in (1.1.9) by the direct sum of functors

$$\Gamma_2^2(G^{ab}) = L(G^{ab},1)_3 \oplus \Gamma(G^{ab}) \otimes \mathbb{Z}/2. \tag{7.1.14}$$

7.1.15. THEOREM: *There is a sequence*

$$0 \to \Gamma_2^2(G^{ab}) \xrightarrow{i} \tilde{\otimes}^3(G,\theta) \xrightarrow{\delta} \tilde{\tilde{\Gamma}}(G,\theta) \xrightarrow{\tilde{\tilde{p}}} \Gamma(G^{ab}) \to 0$$

which is natural for generalized induced maps in (7.1.7) and which satisfies $\delta i = 0$, $\tilde{\tilde{p}}\,\delta = 0$. *Moreover if G is a free group and if $(G,\theta) = (G,\theta_0)$ is given by (7.1.3) then this sequence is exact.*

This result is a consequence of (7.4.9) and (7.5.1) below. The surjection $\tilde{\tilde{p}} = (\gamma p, 0)^\square$ is defined in (7.1.10). The map i in (7.1.15) is given by the coordinates

$$\begin{cases} i_1 & : & L(A,1)_3 \hookrightarrow \otimes^3 A, \text{ with } A = G^{ab}, \\ i_2 & : & \Gamma(A) \otimes \mathbb{Z}/2 \to \phi(A), \\ i_3 & : & L(A,1)_3 \to \phi(A). \end{cases} \tag{7.1.16}$$

Here i_1 is the inclusion as in (6.1.5) and $i_2 = \tau(\tau \otimes \mathbb{Z}/2)$ can be defined by

$$i_2(\gamma a \otimes 1) = aa\hat{\otimes}aa, \ a \in A. \tag{1}$$

Finally the correction term i_3 is given by the formula

$$i_3[[\{x\},\{y\}],\{z\}] = \nabla(x+y,z) - \nabla(x,z) - \nabla(y,z) \tag{2}$$

$$= xy\hat{\otimes}zy + yx\hat{\otimes}zx + T[(xz - zx) \otimes \theta(y) + (yz - zy) \otimes \theta(x)] \tag{3}$$

where $x, y, z \in G$ and where $\{x\} \in G^{ab}$ is the class represented by x. The function ∇ is defined in (7.1.5)(3); it is easy to check that (3) holds and that (3) satisfies the relations of Lie brackets in (2). In particular (3) is linear in x, y and z. As in (6.1.4) we see that $\delta i = 0$ in (7.1.15).

We derive from (7.1.15), (7.1.8) and (7.1.9) the following natural commutative diagram of exact sequences where G is a free group with $A = G^{ab}$ and where (G, θ_0) is given by (7.1.3).

$$(7.1.17) \qquad \begin{array}{ccccccc} L(A,1)_3 & \overset{i}{\rightarrowtail} & \otimes^3 A & \overset{\delta}{\longrightarrow} & \tilde{\Gamma}(G) & \overset{\tilde{p}}{\twoheadrightarrow} & \Gamma(A) \\ \uparrow p & & \uparrow p & & \uparrow h & & \| \\ \Gamma_2^2 A & \overset{i}{\rightarrowtail} & \tilde{\otimes}^3(G,\theta_0) & \overset{\delta}{\longrightarrow} & \tilde{\tilde{\Gamma}}(G,\theta_0) & \overset{\tilde{\tilde{p}}}{\twoheadrightarrow} & \Gamma(A) \\ \downarrow p & & \downarrow Ip & & \downarrow h & & \| \\ (\Gamma A)\otimes \mathbb{Z}/2 & \overset{i}{\rightarrowtail} & \tilde{\otimes}^4 A_2 & \overset{\delta}{\longrightarrow} & \bar{\Gamma}(G, q\theta_0) & \overset{\bar{p}}{\twoheadrightarrow} & \Gamma(A) \end{array}$$

Using the central push outs in (7.1.8) and (7.1.9) we see that the exactness of the row in the middle implies the exactness of the top row and of the bottom row respectively. This yields a proof of theorem (5.1.15) and of theorem (4.1.4), a further proof of the exactness of the top row is given in chapter 6.

We have to check that diagram (7.1.17) commutes. This is clear by (7.1.10) and by the equation $Ii_3 = 0$ where we use (7.1.16)(3) and (5.1.2)(3). In fact, the equation $Ii_3 = 0$ was the original reason to introduce the relations (5.1.2)(2),(3) in the definition of $\tilde{\otimes}^4(A)$. Moreover, we obtain the naturality of diagram (7.1.17) for generalized induced maps since $I\nabla_\eta$ with ∇_η in (7.1.17) is exactly the same as ∇_η in (5.1.12). Also (5.1.13) and (5.1.14) and (5.1.17) are consequences of the corresponding results above, see (7.1.12), (7.1.13) and (7.1.11).

7.2. Crossed square groups

Crossed square groups appear naturally in the definition of tensor products of quadratic chain complexes, see [BCH] and (7.3) below. We here describe basic properties of crossed square groups; the square groups in (7.1) are special crossed square groups with a trivial action.

7.2.1. DEFINITION: We consider pairs (q, h) where $q : G \to \pi$ is a homomorphism between groups and where $h : G \to C$ is a *q-crossed homomorphism*; that is C is a π-module and h satisfies

$$(1) \qquad h(x + y) = h(x)^{q(y)} + h(y) \text{ for } x, y \in G.$$

The tensor product $C \otimes C$ over $\mathbb{Z}$ is a π-module by

$$(2) \qquad (a \otimes b)^{\alpha} = a^{\alpha} \otimes b^{\alpha}$$

for $a, b \in C$, $\alpha \in \pi$. We also consider C as a G-module via q so that we may write for short $a^x = a^{q(x)}$. We call a function

$$(3) \qquad \theta : G \to C \otimes C$$

an *h-square map* if

$$(4) \qquad \theta(x+y) = \theta(x)^y + \theta(y) + h(x)^y \otimes h(y)$$

holds for $x, y \in G$. Here $\theta(x)^y = \theta(x)^{q(y)}$ is given by (2). We call the tuple $(G, \theta) = (G, \pi, C, q, h, \theta)$ a *crossed square group*. In case $\pi = 0$, $C = G^{ab}$ and $h : G \to G^{ab}$ this is the same as a square group, see (7.1.1).

7.2.2. LEMMA: *Let $G = < Z >$ be a free group generated by Z and let a function $f : Z \to C$ be given. Then there is a unique h-square map*

$$\theta = \theta_f : G \to C \otimes C$$

with $\theta(e) = f(e)$ for $e \in Z$.

Proof: We have $h(0) = 0$ and whence $\theta(0) = 0$ for each h-square map θ. Moreover we get $\theta(x - x) = \theta(0) = 0$ so that

$$(1) \qquad \begin{aligned} \theta(-x) &= -\theta(x)^{-x} - h(x)^{-x} \otimes h(-x) \\ &= -\theta(x)^{-x} + h(x)^{-x} \otimes h(x)^{-x} \\ &= (-\theta(x) + h(x) \otimes h(x))^{-x}. \end{aligned}$$

Let $(Mon(\pm Z), +)$ be the free monoid generated by the elements e and $-e$, $e \in Z$ and let $q : Mon(\pm Z) \twoheadrightarrow < Z > = G$ be the canonical surjection. We define a function

$$(2) \qquad \theta : Mon(\pm Z) \to C \otimes C$$

on generators by $\theta(e) = f(e)$ and $\theta(-e) = (-f(e) + h(e) \otimes h(e))^{-e}$ for $e \in Z$.

Moreover we assume that θ in (2) satisfies the equation (7.2.1)(4). Then θ in (2) is well defined since (7.2.1)(4) yields the following equations.

$$\begin{aligned}\theta((x+y)+z) &= \theta(x+y)^z + \theta(z) + h(x+y)^z \otimes h(z)\\ &= \theta(x)^{y+z} + \theta(y)^z + h(x)^{y+z} \otimes h(y)^z + \theta(z)\\ &\quad + h(x)^{y+z} \otimes h(z) + h(y)^z \otimes h(z),\end{aligned}$$

$$\begin{aligned}\theta(x+(y+z)) &= \theta(x)^{y+z} + \theta(y+z) + h(x)^{y+z} \otimes h(y+z)\\ &= \theta(x)^{y+z} + \theta(y)^z + \theta(z) + h(y)^z \otimes h(z)\\ &\quad + h(x)^{y+z} \otimes h(y)^z + h(x)^{y+z} \otimes h(z).\end{aligned}$$

Now it is clear by definition of $\theta(-e)$, $e \in Z$, that θ in (2) satisfies $\theta(e-e) = \theta(-e+e) = 0$. This shows that θ yields a unique factorization $\theta_f : G \to C \otimes C$ with $\theta_f q = \theta$. □

7.2.3. EXAMPLE: Let $G =< Z >$ be a free group and let $q : G \to \pi$ be a homomorphism. Then we have the universal q-crossed homomorphism

$$h : G \to C(q) = \bigoplus_Z \mathbb{Z}[\pi] \tag{1}$$

when $C(q)$ is the free π-module generated by Z and where h is defined by $h(e) = e$ for $e \in Z$. Let

$$\theta_0 : G \to C(q) \otimes C(q) \tag{2}$$

be the unique h-square map with $\theta_0(e) = 0$ for $e \in Z$, see (7.2.2). Thus we get the *crossed square group*

$$(G, \theta_0) = (G, \pi, C(q), q, h, \theta_0) \tag{3}$$

which clearly depends on the choice of generators Z in G. Using the augmentation $\epsilon : \mathbb{Z}[\pi] \to \mathbb{Z}$ one gets the projection

$$\epsilon : C(q) \to C(q) \otimes_\epsilon \mathbb{Z} = G^{ab} \tag{4}$$

for which $(G, (\epsilon \otimes \epsilon)\theta_0) = (G, \epsilon_* \theta_0)$ coincides with the square group in (7.1.3).

7.2.4. DEFINITION: We define the *square derivation* θ on the following category $\mathbf{G}$. Objects in $\mathbf{G}$ are homomorphisms $q : G \to \pi$ where $G =< Z >$ is a free group. Morphisms $q' \to q$ in $\mathbf{G}$ are pairs (g, φ) of homomorphisms $g : G' \to G$, $\varphi : \pi' \to \pi$ with $qg = \varphi g'$. Such a morphism induces a unique φ-equivariant homomorphism

$$(1) \qquad (g, \varphi)_* : C(q') \to C(q)$$

with $(g, \varphi)_* h = hg$. Whence C in (7.2.3)(1) is a functor (compare the notation in (6.2)):

$$(2) \qquad C : \mathbf{G} \to \mathbf{Mod}_{\mathbb{Z}}^{\wedge}.$$

Using the functor we obtain a natural system H on $\mathbf{G}$ defined by

$$(3) \qquad H(g, \varphi) = Hom_{\varphi}(C(q'), C(q) \otimes C(q))$$

where the right hand side is the abelian group of φ-equivariant homomorphisms. Using the square maps θ_0 in (7.2.3) we define the φ-equivariant homomorphism

$$(4) \qquad \begin{cases} \theta(g, \varphi) \in H(g, \varphi) \text{ by} \\ \theta(g, \varphi)(e') = \theta_0(g(e')) \text{ for } e' \in Z' \end{cases}$$

where Z' is a basis of $G' =< Z' >$. The operator θ in (4) can also be described as follows. Consider the diagram

$$(5) \qquad \begin{array}{ccccc} G & & \longrightarrow & & C(q) \otimes C(q) \\ \uparrow g & & & \nearrow \theta(g,\varphi) & \uparrow \otimes^2 (g,\varphi)_* \\ & & C(q') & & \\ & \nearrow h & & & \\ G' & & \longrightarrow & & C(q') \otimes C(q') \end{array}$$

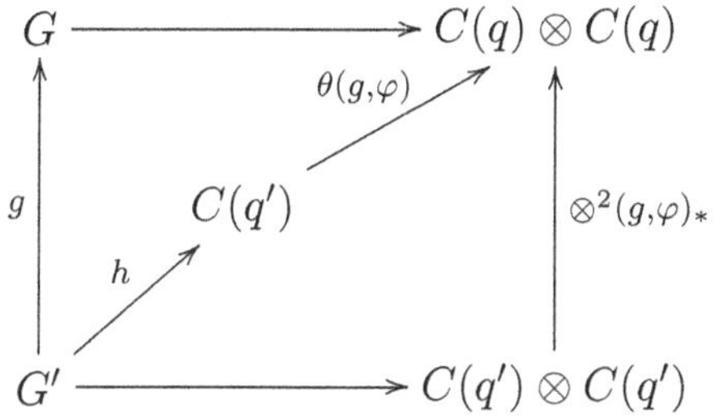

This diagram does not commute. The difference $-(g, \varphi)_*^{\otimes 2} \theta_0 + \theta_0 g$, however, is a $\varphi q'$-crossed homomorphism so that one gets a unique φ-equivariant homomorphism $\theta(g, \varphi)$ as in (5) with

(6) $$\theta(g,\varphi)h = -(g,\varphi)_*^{\otimes 2}\theta_0 + \theta_0 g.$$

Using (6) one readily sees that the operator $\theta : \mathbf{G} \to H$ which carries (g,φ) to $\theta(g,\varphi) \in H(g,\varphi)$ is a derivation, that is

(7) $$\theta(gg',\varphi\varphi') = (\otimes^2(g,\varphi)_*)\theta(g',\varphi') + \theta(g,\varphi)(g',\varphi')_*.$$

We call θ the *square derivation* on the category $\mathbf{G}$. This derivation represents the cohomology class

(8) $$\{\theta\} \in H^1(\mathbf{G}, H),$$

see [BAH], which we call the *square class* on $\mathbf{G}$. One can check that the square class does not depend on the choice of the basis Z in the free group G, in fact, the class $\{\theta\}$ can be obtained by replacing θ_0 in (4) by any h-square map θ_1. The difference $\theta_0 - \theta_1$ then is an h-crossed homomorphism $G \to C(q) \otimes C(q)$, see (7.2.1)(4), which yields a homomorphism $C(q) \to C(q) \otimes C(q)$ of π-modules. Whence the derivations given by θ_0 and θ_1 respectively differ only by an inner derivation.

7.3. The quadratic chain complex of the James construction

Using tensor products of quadratic chain complexes we get the quadratic chain complex of the James construction JX of a CW-complex X. In case $X = K(G,1)$ is a one point union of 1-spheres this gives us the quadratic James construction $J_Q G$ of a free group $G = \pi_1 X$. Here $J_Q G$ is not a functor in G but canonically a pseudo–functor in G. Recall that each CW-complex X with $X^0 = *$ yields a quadratic chain complex $\sigma = \sigma(X)$ given by the diagram

(7.3.1)
$$\begin{array}{ccccccc} & & & C_2 \otimes C_2 & & & \\ & & & \downarrow w & & & \\ \longrightarrow \sigma_4 & \xrightarrow{d} & & \sigma_3 & \xrightarrow{d} & \sigma_2 & \xrightarrow{d} \sigma_1 \end{array}$$

where w is the quadratic map. We have the quotient map $q : \sigma \to \rho = \lambda(\sigma)$ by dividing out the image of w and the image of the Pfeiffer commutator map $w = dw$, that is $\rho_3 = \sigma_3 / w(\otimes^2 C_2)$ and $\rho_2 = \sigma_2 / dw(\otimes^2 C_2)$, compare IV.3.3 in [BCH]. This quotient is a crossed chain complex which is $\rho(X)$ in

(6.2.1) for $\sigma = \sigma(X)$. Moreover we have the chain complex $C(\rho) = C$ together with a map $h : \rho \to C(\rho)$. Here $h_1 : \rho_1 \to C_1$ is the universal q-crossed homomorphism with respect to the quotient map $q : \rho_1 \to \pi_1 = \rho_1/d\rho_2$ and $h_2 : \rho_2 \to C_2 = \rho_2^{ab}$ is the abelianization. Moreover h is the identity in degree ≥ 3, compare III.2.1 in [BCH]. The chain complex $C = C_*\hat{X}$ is the cellular chain complex of the universal covering $\hat{X}$ for $\rho = \rho(X)$. Eachof the objects $\sigma = \sigma(X)$, $\rho = \lambda\sigma = \rho(X)$ and $C = C(\rho) = C_*(\hat{X})$ respectively is totally free with the basis in degree n given by the set of n-cells of X. We need the following notation, compare [FM].

7.3.2. DEFINITION: Recall that a track category **TC** is the same as a groupoid enriched category, see VI.3.1 in [BCH]. We write the composition of tracks additively. Let **E** be a category. A *pseudo functor*

$$(F, \Theta) : \mathbf{E} \to TC \tag{1}$$

carries a morphism $f : X \to Y$ in **E** to a morphism $F(f) : FX \to FY$ in **C** with $F(1) = 1$ for the identity 1. Moreover Θ carries a pair of morphisms $f : X \to Y$, $g : X' \to X$ to a track

$$\Theta(f, g) : F(fg) \simeq F(f)F(g) \tag{2}$$

which is the trivial track if $g = 1$ or $f = 1$. Moreover the associativity law

$$\Theta(f, gh) + F(f)_*\Theta(g, h) = \Theta(fg, h) + (Fh)^*\Theta(f, g) \tag{3}$$

is satisfied. In case $\Theta = 0$ is the trivial track for all f, g we see that F is the same as a functor $\mathbf{E} \to \mathbf{C}$. A *natural transformation*

$$(\alpha, A) : (F, \Theta) \to (F', \Theta') \tag{4}$$

between pseudo functors caries an object X to a morphism $\alpha_X : FX \to F'X$ and a morphism $f : X \to Y$ to a track $A(f) : F'(f)\alpha_X \simeq \alpha_Y F(f)$ with $A(1) = 0$ and

$$A(fg) = (Fg)^*A(f) + (F'f)_*A(g). \tag{5}$$

Let $\mathbf{Q}$ be the category of totally free quadratic chain complexes and let $T_0\mathbf{Q}$ be the corresponding track category with tracks given by 0-homotopies, see IV.4.8 in [BCH]. For maps $f, g : \sigma' \to \sigma$ in $\mathbf{Q}$ such a zero homotopy is a $\pi_1(f)$-equivariant homomorphism

$$\alpha_2 : C_2' \to C_2 \otimes C_2 = \otimes^2 C_2 \tag{7.3.3}$$

with $-f_2 + g_2 = dw\alpha_2 q$, $-f_3 + g_3 = w\alpha_2 qd$ and $-f_n + g_n = 0$ for $n = 1, n \geq 3$.

The *tensor product* of quadratic chain complexes is a pseudo functor

$$(\otimes, \Theta) : \mathbf{Q} \times \mathbf{Q} \to T_0\mathbf{Q}. \tag{7.3.4}$$

This pseudo functor carries the pair (A, B) of quadratic chain complexes to the tensor product $A \otimes B$ defined in IV.12.2 of [BCH]. Moreover (7.3.4) carries a morphism $F = (F^A, F^B) : (A', B') \to (A, B)$ in $\mathbf{Q} \times \mathbf{Q}$ to the induced homomorphism

$$F^A \otimes F^B = (F^A \otimes 1)(1 \otimes F^B) \tag{1}$$

defined in IV.12.10 of [BCH]. For $G = (G^A, G^B) : (A'', B'') \to (A', B')$ the track

$$\Theta(F, G) : (F^A G^A) \otimes (F^B G^B) \overset{\circ}{\simeq} (F^A \otimes F^B)(G^A \otimes G^B) \tag{2}$$

in $T_0\mathbf{Q}$ is given by a 0-homotopy

$$\Theta(F, G) : (C^{A''} \otimes C^{B''})_2 \to \otimes^2 (C^A \otimes C^B)_2 \tag{3}$$

where $C^A \otimes C^B = C\lambda(A \otimes B)$ with $C^A = C\lambda(A)$, $C^B = C\lambda(B)$. Using the square derivation θ in (7.2.4) we obtain homomorphisms

$$\begin{cases} \theta(F_1^B) & : \quad C_1^{B'} \to \otimes^2 C_1^B, \\ \theta(G_1^A) & : \quad C_1^{A''} \to \otimes^2 C_1^{A'} \end{cases} \tag{4}$$

which yield $\Theta(F, G)$ in (3) by the composition

$$\Theta(F, G) = -T\{[(\otimes^2 F_*^A)\theta(G_1^A)] \otimes [\theta(F_1^B) G_*^B]\} \tag{5}$$

where $F_*^A : C_1^{A'} \to C_1^A$, resp. $G_*^B : C_1^{B''} \to C_1^{B'}$, are given by $C\lambda(F^A)$, resp. $C\lambda(G^B)$, or equivalently by $C(F_1^A)$, resp. $C(G_1^B)$, where we use the functor C in (7.2.4)(2). Moreover T in (5) is the permutation

$$(6) \qquad \begin{cases} T : \otimes^2 C_1^A \otimes \otimes^2 C_1^B \to \otimes^2 (C_1^A \otimes C_1^B), \\ T(a_1 \otimes a_2 \otimes b_1 \otimes b_2) = a_2 \otimes b_2 \otimes a_1 \otimes b_1. \end{cases}$$

Equation (5) means that the right hand side of (5) is the only non trivial component of the homomorphism $\Theta(F, G)$ in (3). The property that θ is a derivation, see (7.2.4), corresponds exactly to the property (7.3.2)(3) of the pseudo functor $(\otimes, \Theta)$. We obtain (5) by IV.12.4 and IV.12.11 in [BCH]. Theorem IV.12.1 of [BCH] shows that one has the isomorphism

$$(7.3.5) \qquad \sigma(X \times Y) \simeq \sigma X \otimes \sigma Y$$

where X, Y are CW-complexes with $X^0 = * = Y^0$. Moreover this isomorphism is natural if we use the pseudo functor $(\otimes, \Theta)$ in (7.3.4). This naturality is obtained by the fact that for 1–dimensional CW-complexes X, Y the 0-homotopy (7.3.4)(2) is unique since in this case $\pi_3(X \times Y) = 0$. More precisely one has in this case a natural transformation of pseudo functors in (X, Y)

$$(1) \qquad (\sigma X \otimes \sigma Y, \Theta) \xrightarrow{(b,B)} (\sigma_S(X \times Y), 0)$$

which is a weak equivalence. Here σ_S is the model functor in IV.6.7 of [BCH] and $\sigma X = \pi_1 X$, $\sigma Y = \pi_1 Y$ are quadratic chain complexes concentrated in degree 1. The natural map (1) is the model map b in diagram IV.6.8(6) of [BCH], the track B in (1) is given by a homotopy for this diagram considered as a diagram in $Fil\,\mathbf{Q}$.

For the James construction IV.A.18 in [BCH] of totally freee quadratic chain complexes we have the isomorphism

$$(2) \qquad J_Q(\sigma X) \cong \sigma J(X).$$

Again the isomorphism is natural for cellular maps $X \to Y$ if we use the structure of J_Q as a pseudo functor. Recall that $J_Q(\sigma)$ is a totally free quadratic chain complex with a basis given by the free monoid $Mon(Z_*)$ where Z_* is a basis of σ. The boundary in degree ≤ 3 of $J_Q(\sigma)$ can be computed by the canonical map

$$(3) \qquad \theta : (\sigma \otimes \sigma) \otimes \sigma \to J_Q(\sigma)$$

which is surjective in degree ≤ 3 and which is defined on generators by

$$\begin{aligned} e_1 \times e_2 \times e_3 &\mapsto e_1 e_2 e_3, \\ e_1 \times e_2 \times *,\ e_1 \times * \times e_2,\ * \times e_1 \times e_2 &\mapsto e_1 e_2, \\ e_1 \times * \times *,\ * \times e_1 \times *,\ * \times * \times e_1 &\mapsto e_1, \end{aligned}$$

where $e_1, e_2, e_3 \in Z_*$. Using the map v in (3) we obtain by (7.3.4) the pseudo functor

$$(4) \qquad (J_Q, \Theta) : \mathbf{Q} \to T_0\mathbf{Q}.$$

Here the 0-homotopy Θ is defined in the same way as in (7.3.4)(5) simply by setting $A = B$, $F^A = F^B$, $G^A = G^B$. Using this pseudo functor the isomorphism (2) above is natural. This means, in particular, that we get by (1) theorem (7.3.6) below which is crucial for the main result of this book. For this we consider a free group G as a totally free quadratic chain complex $\sigma = G$ concentrated in degree 1. Therefore we obtain by (3) the *quadratic James construction $J_Q(G)$ of the free group G*. Let $\mathbf{Gr}_c$ be the category of free groups considered as a subcategory of $\mathbf{Q}$. Then the restriction of (4) gives the pseudo functor

$$(5) \qquad (J_Q, \Theta) : \mathbf{Gr}_c \to T_0\mathbf{Q}.$$

The totally free quadratic chain complex $J_Q(G)$ with $G = < Z >$ has the free monoid $Mon(Z)$ as a basis. Moreover we have the natural quoteint map

$$(6) \qquad p : J_Q(G) \to J(G) = \lambda J_Q(G)$$

which is the identity on the basis $Mon(Z)$. Here the crossed chain complex JG is explicitly described in (6.2.2). In a similar way one has an explicit description of $J_Q(G)$ by using (3) and the formulas for the tensor product in IV.12.2 of [BCH]. Since this description is lengthy we do not recall all formulas here and refer the reader to [BCH].

We now consider the case when X in (2) is a one point union

$$X = K(G,1) = M_A = \bigvee_Z S^1$$

of 1-spheres as in (6.2.4) where $G = \pi_1 X$, $A = G^{ab}$. The quadratic James construction of the free group G then has the following property which corresponds to the property of the crossed James construction in (6.2.5).

7.3.6. THEOREM: *For a free group $G =< Z >$ there is an isomorphism*

$$\psi : \sigma JK(G,1) \cong J_Q(G)$$

of quadratic chain complexes. Moreover there is a natural transformation of pseudo functors in **G**

$$b : (J_Q(G), \Theta) \to (\sigma_S JK(G,1), 0)$$

where σ_S again is the model functor in IV.6.7 of [BCH] and where b is a weak equivalence as in (7.3.5)(1).

As in (6.2.5) the isomorphism ψ carries the product cell $e_1 \times \cdots \times e_n$ to the word $e_1 \ldots e_n$ where we identify a 1–cell e in $K(G,1) = \bigvee_Z S^1$ with the corresponding generator $e \in Z$ in G. The theorem is a consequence of (7.3.5)(1), see also IV.A.18(6) in [BCH].

7.4. The quadratic James construction of a free group

We study algebraic properties of the quadratic James construction of a free group G. Let $G =< Z >$ be a free group and let $J_Q(G)$ be the quadratic James construction of G described in (7.3.5)(5). We clearly have $(J_Q G)_1 = G$. Moreover there is a commutative diagram of G-groups

(7.4.1)

$$\begin{array}{ccccccccc}
 & & & & C_2 \otimes C_2 & & & & \\
 & & & & \downarrow w & & & & \\
\longrightarrow & (J_QG)_4 & \longrightarrow & & (J_QG)_3 & \xrightarrow{d} & (J_QG)_2 & \xrightarrow{d} & G \\
 & \| & & & \downarrow p & & \downarrow p & & \| \\
\longrightarrow & (JG)_4 & \longrightarrow & & (JG)_3 & \xrightarrow{d} & (JG)_2 & \xrightarrow{d} & G \\
 & \downarrow & & & \downarrow q & & \downarrow q & & \| \\
\longrightarrow & \otimes^4 G^{ab} & \xrightarrow{0} & & \otimes^3 G^{ab} & \xrightarrow{0} & \otimes^2 G^{ab} & \xrightarrow{0} & G^{ab}
\end{array}$$

which extends diagram (6.2.7). Here w is the *quadratic map* of J_QG with

(7.4.2)
$$C_2 = (JG)_2^{ab} = C_1 \otimes C_1$$

where $C_1 = C(q)$ is given by the abelianization $q : G \to G^{ab}$, see (7.2.3)(1). The cokernel of w is $(JG)_3$ and the cokernel of $w = dw$ is $(JG)_2$; the map p in (7.4.1) is the quoteint map which is also the map in (7.3.5)(6) above.

We now construct the following natural commutative diagram where v is given by (7.3.5)(3); this is the quadratic analogue of diagram (6.2.8).

(7.4.3)

$$\begin{array}{ccccccc}
((G \otimes G) \otimes G)_3 & & G & & (G \otimes G)_2 & & \\
\downarrow v & & \downarrow f & & \downarrow v & & \\
(J_QG)_3 & \xrightarrow{d} & \hat{J}_QG & \rightarrowtail & (J_QG)_2 & \xrightarrow{d} & G \\
\downarrow h_3 & (i) & \downarrow h_2 & (ii) & \downarrow h_2 & & \downarrow h_1 \\
\tilde{\otimes}^3 G & \xrightarrow[\delta]{} & \tilde{\tilde{J}}G & \rightarrowtail & (\tilde{\tilde{J}}G)_2 & \xrightarrow{\tilde{\tilde{d}}} & G
\end{array}$$

Here $\hat{J}_QG$ and $\tilde{\tilde{J}}G$ denote the kernels of d and $\tilde{\tilde{d}}$ respectively. The function f is defined by

$$f(x) = v(x \mathbin{\bar{\bar{\otimes}}} x)$$

so that $pf : G \to \hat{J}G$ coincides with f in (6.2.8).

7.4.4. CONSTRUCTION (OF h_3): Recall that we have for the free group G the square group (G, θ_0) in (7.1.2) for which we set

$$\text{(1)} \qquad \begin{cases} \tilde{\otimes}^3(G) &= \tilde{\otimes}^3(G, \theta_0) = \otimes^3 A \oplus \phi(A), \\ \phi(A) &= (\otimes^2 A)\hat{\otimes}(\otimes^2 A) \otimes \mathbb{Z}/2 \end{cases}$$

with $A = G^{ab}$, see (7.1.4). We now define the map

$$\text{(2)} \qquad h_3 = (qp, r) : (J_Q G)_3 \to \tilde{\otimes}^3(G)$$

by the components qp and r where qp is the map in (7.4.1) and where r is constructed as follows. For C_2 in (7.4.2) we have the map

$$\text{(3)} \qquad q = \epsilon \otimes \epsilon : C_2 = C_1 \otimes C_1 \to A \otimes A$$

which is induced by q in (7.4.1) or by ϵ in (7.2.3)(4). We now consider the folowing push out diagram from IV.3.4(1) of [BCH] for $\sigma = J_Q G$, $\rho = \lambda(\sigma) = JG$, $C = C(\rho)$.

$$\text{(4)} \qquad \begin{array}{ccccc} J & \xrightarrow{(d,1)_*} & C_2 \otimes C_2 & \xrightarrow{q \otimes q} & \Phi(A) \\ \downarrow & & \downarrow w & \nearrow r & \\ (E_3 \vee \sigma_2)_2 & \xrightarrow{\nu} & \sigma_3 & & \\ \downarrow & & \downarrow p & & \\ C_3 & \xleftarrow{\cong} & \rho_3 & & \end{array}$$

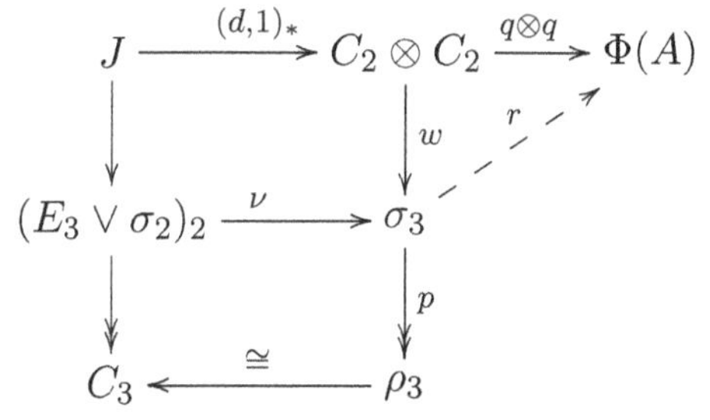

Here $J = C_3 \otimes C_3 \oplus C_3 \otimes C_2 \oplus C_2 \otimes C_3$ is given by C and q is the map in (3). Since $qd = 0$ by (7.4.1) we see that the composition of maps in the top row of (4) is trivial. Therefore there is a unique homomorphism r in (4) with $r\nu = 0$ and $rw = q \otimes q$. This completes the definition of the component r in (2). All maps in (4) are equivariant with respect to the action of G so that h_3 is also G-equivariant where G acts trivially on $\tilde{\otimes}^3(G)$. The naturality of the surjection h_3 is the reason for the definition of the induced maps for $\tilde{\otimes}^3$ in (7.1.4). This is proved by the following lemma.

7.4.5. LEMMA: *The map h_3 is natural.*

Proof: For a homomorphism $\eta : G' \to G$ the induced map $\eta_3 = J_Q(\eta)_3$ is given by the commutative diagram (see (7.3.5)(3))

(1)
$$\begin{array}{ccc} ((G' \otimes G') \otimes G')_3 & \xrightarrow{v} & J_Q(G')_3 = \sigma'_3 \\ \downarrow{\scriptstyle (\eta\otimes\eta)\otimes\eta} & & \downarrow{\scriptstyle \eta_3} \\ ((G \otimes G) \otimes G)_3 & \xrightarrow[v]{} & J_Q(G)_3 = \sigma_3 \end{array}$$

Here we have by (7.3.4)(1) the equation

$$(\eta \otimes \eta) \otimes \eta = (\eta \otimes 1) \otimes 1 \circ (1 \otimes \eta) \otimes 1 \circ (1 \otimes 1) \otimes \eta$$

which shows that for a generator $efg = v((e \times f) \times g)$, $e, f, g \in Z$, we get

(2)
$$\eta_3(efg) = v(\eta e \overset{=}{\otimes} \eta f)\bar{\otimes}\eta g,$$

compare the definition in IV.12.10 of [BCH]. We now consider the diagram

(3)
$$\begin{array}{ccccc} \sigma'_3 & \xrightarrow{(q',r)} & \rho'_3 \oplus \Phi(A') & \xrightarrow{q\otimes 1} & \otimes^3 A \oplus \Phi(A') \\ \downarrow{\scriptstyle \eta_3} & & \downarrow{\scriptstyle \eta_\#} & & \downarrow{\scriptstyle \eta_*} \\ \sigma_3 & \longrightarrow & \rho_3 \oplus \Phi(A) & \longrightarrow & \otimes^3 A \oplus \Phi(A) \end{array}$$

where the top and the bottom row describe the map h_3. Since $q \otimes q$ in (7.4.4)(4) is natural there exists a map $\eta_\#$ for which (3) commutes. Moreover since the action of G on $\Phi(A)$ is trivial the coordinate $\rho'_3 \to \Phi(A)$ of $\eta_\#$ admits a factorization $\rho'_3 \to \otimes^3 A \to \Phi(A)$. Whence there exists a map $\eta_* = \tilde{\otimes}^3(\eta)$ for which (3) commutes. We now compute the coordinate $\otimes^3 A \to \Phi(A)$ of η_*. For this we define the function

(4)
$$\begin{cases} R : G \times G \times G \to \Phi(A), \\ R(x, y, z) = rv((x \overset{=}{\otimes} y)\bar{\otimes} z) \end{cases}$$

where r is the function in (7.4.4)(4). We claim the formula

(5)
$$R(x, y, z) = T((x \otimes y - y \otimes x) \otimes \theta_0(z))$$

is satisfied. We prove (5) in (6) below. Equation (5) implies by (2) that the coordinate $\otimes^3 A \to \Phi(A)$ of η_* in (3) is given by $\theta_\#(\eta)$, compare (7.1.4). This completes the proof that h_3 is natural. □

(6)*Proof* (of(5)): Since a generator efg is an element of E_3 in (7.4.4)(4) we see that $R(e,f,g) = 0$. Moreover one can check that

(a) $R(x,y,z)$ is linear in x and y and
(b) $R(x,y,z+z') = R(x,y,z) + R(x,y,z') - y \otimes z' \hat{\otimes} x \otimes z + x \otimes z' \hat{\otimes} y \otimes z$.

This implies that $R(x,y,g) = 0$ for all x,y and generators g. Using the definition of T in (7.1.4)(5) we see that the function on the right hand side of (5) has the same properties. This implies that equation (5) holds. The proof of (a) and (b) is an application of the distributivity laws for $\bar{\bar{\otimes}}$ and $\bar{\otimes}$ in (2), compare the formulas IV.12.5 in [BCH]. □

7.4.6. CONSTRUCTION (OF h_2): We now define h_2 and the bottom row of diagram (7.4.3) by the use of the following commutative diagram in which all maps are σ_2-equivariant homomorphisms, $\sigma_n = (J_Q G)_n$, $n \geq 1$.

(1)

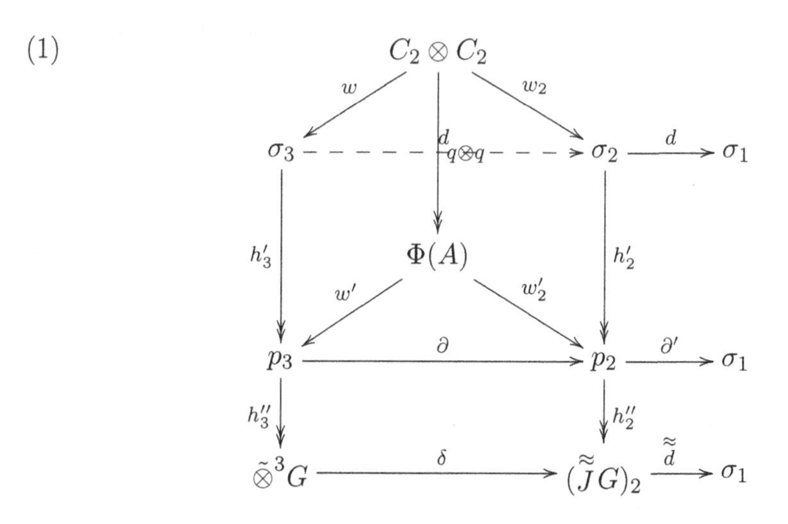

Each square of unbroken arrows in this diagram is a central push out diagram, this defines P_2, P_3 and $(\tilde{\tilde{J}} G)_3$. For this we use the fact that the maps w and $w_2 = dw$ in a quadratic chain complex are central. Also the map ∂ induced by $d : \sigma_3 \to \sigma_2$ is central. This is seen as follows.

Let $x \in P_3$, $y \in P_2$, $x = h'_3 x'$, $y = h'_2 y'$. Then we have the Pfeiffer commutator

$$(2) \qquad -y' - dx' + y' + (dx')^{dy'} = < y', dx' > = w_2(\{y'\} \otimes \{dx'\})$$

Here $\{y'\} \otimes \{dx'\} \in C_2 \otimes C_2$ lies in the kernel of $q \otimes q$. Moreover we have

$$(3) \qquad h'_2(dx')^{dy'} = \partial(h'_3 x')^{dy'} = \partial h'_3 x' = \partial x$$

since dy' acts trivially on P_3. In fact P_3 is $C_3 \oplus (\otimes^2 A)\hat{\otimes}(\otimes^2 A)$ by (7.4.4)(4). Now (3) and (2) show

$$(4) \qquad -y - \partial x + y + \partial x = h'_2 < y', dx' > = 0$$

so that ∂ is indeed central. Therefore the bottom row of (1) is well defined. For this ∂' is induced by $d : \sigma_2 \to \sigma_1$ since $dw_2 = 0$ and $\overset{\approx}{d}$ is induced by ∂' since $\partial\partial = 0$. Moreover we obtain the maps h_2, h_3 in (7.4.3) by the composition of vertical arrows in (1), that is $h_2 = h''_2 h'_2$, $h_3 = h''_3 h'_3$. Since all maps in (1) are natural with respect to homomorphisms $\eta : G' \to G$ we see that η induces a homomorphism

$$(5) \qquad \overset{\approx}{J}(\eta) : \overset{\approx}{J}(G') \to \overset{\approx}{J}(G)$$

where we define $\overset{\approx}{F}(G) = kernel(\overset{\approx}{d})$, see (7.4.3). Since, however, J_Q is only a pseudo functor the induced maps (5) satisfy only a composition law as in (7.1.11), this follows from (7.3.4)(2).

By (7.4.6) the construction of diagram (7.4.3) is complete. We now consider the homology groups of the rows of diagram (7.4.3). Let

$$\pi_n(\sigma) = kernel(d : \sigma_n \to \sigma_{n-1})/image(d : \sigma_{n+1} \to \sigma_n)$$

be the homology of a quadratic chain complex $\sigma = (\cdots \to \sigma_2 \to \sigma_1, w)$. Then we have for a CW-complex Y with T^0 the natural isomorphism

$$(7.4.7) \qquad \pi_n(\sigma Y) \cong \pi_n Y \text{ for } n \leq 3,$$

compare IV.7.5 of [BCH]. This implies the next result which is the analogue of (6.2.6).

7.4.8. PROPOSITION: *Let G be a free group with $A = G^{ab}$. Then one has natural isomorphisms*

$$\begin{aligned}\pi_1 J_Q(G) &= A,\\ \pi_2 J_Q(G) &= \Gamma A,\\ \pi_3 J_Q(G) &= \Gamma_2^2 A.\end{aligned}$$

Proof: We apply (7.4.7) for $Y = JK(G,1) = JM_A \simeq \Omega M(A,2)$ so that (7.4.8) follows from (1.1.9). The isomorphisms in (7.4.8) are induced by ψ in (7.3.6) and are mapped via the functor λ to the corresponding isomorphisms in (6.2.6). □

The next corollary is the analogue of (6.2.9).

7.4.9. COROLLARY: *For a free group G one has the natural exact sequence*

$$\Gamma_2^2 G^{ab} \overset{i}{\rightarrowtail} \tilde{\otimes}^3 G \xrightarrow{\delta} \approx{J} G \overset{\approx{p}}{\twoheadrightarrow} \Gamma(G^{ab}).$$

Proof: We have to show that the map $h = (h_3, h_2, h_1)$ in (7.4.3) is a 3-equivalence, that is $\pi_i(h)$ is an isomorphism for $i \leq 3$. This is clear for $i = 1$. Let $\approx{J}$ be the bottom row of (7.4.3) and let $\tilde{J}$ be the bottom row of (6.2.8). Then we have the projection $p' : \approx{J} \to \tilde{J}$ induced by $p : J_Q G \to JG$ in (7.4.1). The composition $p'h$ induces an isomorphism $\pi_2(p'h)$ by (6.2.9) and (7.4.8). Whence $\pi_2 h$ is injective. Moreover $\pi_2 h$ is surjective since fx in (7.4.3) corresponds to f in (6.2.8) and hence to $\gamma : G^{ab} \to \Gamma(G^{ab})$. It remains to check that $\pi_3 h$ is an isomorphism. For this we use the fact that $w_2 = dw$ in (7.4.6)(1) or (7.4.1) satisfies

$$kernel(w_2) = \tau\Gamma(K)$$

with $K = kernel(d : C_2 \to C_1)$, compare IV.1.8 in [BCH]. The map $q : C_2 \to \otimes^2 A$ in (7.4.4)(3) maps K surjectively to $\tau\Gamma A$, see (6.4.1)(1). Therefore $q \otimes q$ in (7.4.6)(1) maps $kernel(w_2)$ surjectively to $\tau(\tau \otimes \mathbb{Z}/2)(\Gamma A \otimes \mathbb{Z}/2)$.

Now Whitehead's exact sequence shows that $\pi_3 h$ is an isomorphism. For this see (6.2.6)(2) and IV.3.7, IV.7.3 on [BCH]. Here we also use the properties of central push outs in I.4.20 of [BCH] which show that the kernel of w_2' in (7.4.6)(1) is $\Gamma A \otimes \mathbb{Z}/2$. Since w' is injective this kernel injects to $kernel(\partial)$ and hence injects to $kernel(\delta)$. Now II.2.9 in [BCH] shows that $\pi_3 h$ is an isomorphism. □

7.5. The isomorphism $\tilde{\tilde{J}}G = \tilde{\tilde{\Gamma}}(G,\theta_0)$ for a free group G

In this section we proceed similarly as in (6.3). We consider the pair of maps

$$\tilde{\otimes}^3(G) \xrightarrow{\delta} \tilde{\tilde{J}}G \xleftarrow{h_2 f} G$$

defined by (7.4.3) and we show that this is a (G,θ_0)-quadratic–pair in the sense of (7.1.5). Here we use the square group (G,θ_0) in (7.1.2) which is given by the free group $G =< Z >$. The properties of the pair $(h_2 f, \delta)$ led us originally to the definition of quadratic pairs for square groups in (7.1). Moreover we show

7.5.1. THEOREM: *For a free group $G =< Z >$ the (G,θ_0)-quadratic pair $(h_2 f, \delta)$ above is universal so that $\tilde{\tilde{J}}G = \tilde{\tilde{\Gamma}}(G,\theta_0)$. Moreover this isomorphism is natural for homomorphisms $G' \to G$ between free groups.*

For the proof of (7.5.1) we need the following auxiliary notion; the proof of (7.5.1) is given in (7.5.10) below.

7.5.2. DEFINITION: Let $(G,\theta) = (G,\pi,C,q,h,\theta)$ be a crossed square group. For $x,y,z,z' \in G$ we write for short

$$x^z y^{z'} = (hx)^z \otimes h(y)^{z'} \quad \in C \otimes C \tag{1}$$

and $(xy)^z = x^z y^z$. We define elements $U'(x,y)$, $U_1'(x,y,z)$ and $U_2'(x,y,z)$ in $\otimes^4 C$ as follows.

$$\begin{aligned} U'(x,y) &= T(x^y y \otimes \theta y) + T(x^y y \otimes \theta x)^y \\ &\quad -yy \otimes (xy)^y + (xx)^{2y} \otimes (yx)^y. \end{aligned} \tag{2}$$

Here $T : \otimes^4 C \to \otimes^4 C$ is the permutation in (7.3.4)(6) with $T(x_1x_2 \otimes y_1y_2) = x_2y_2 \otimes x_1y_1$.

$$U_1'(x,y,z) = T(x^y y \otimes \theta z) + (yz + zy) \otimes (zx)^y, \tag{3}$$

$$U_2'(x,y,z) = T(y^z z \otimes \theta x) - (xy + yx)^z \otimes xz. \tag{4}$$

If $\pi = 0$ these elements correspond to the elements $U(x,y)$, $U_1(x,y,z)$ and $U_2(x,y,z)$ in (7.1.5) above.

7.5.3. DEFINITION: Let $(G,\theta) = (G,\pi,C,q,h,\theta)$ be a crossed square group and let K be a G-group. We say that a pair (f,w) of functions

$$G \xrightarrow{f} K \xleftarrow{w} \otimes^4 C$$

is (G,θ)-*strongly quadratic* if the following properties are satisfied. The map w is a G-equivariant homomorphism which maps to the center of K. Moreover let $\mid x,y \mid \in K$ for $x,y \in G$ be defined by the equation

$$f(x+y) = f(y) + \mid x,y \mid^y + f(x)^{2y} + wU'(x,y) \tag{1}$$

where $U'(x,y)$ is given by (7.5.2)(2). Then the functions f and w satisfy the following equations:

$$f(-x) = f(x)^{-2x} + wT(xx \otimes \theta_0(x)^x)^{-2x}, \tag{2}$$

$$-f(x) - f(y) + f(x) + f(y) = w(xy \otimes xy), \tag{3}$$

$$\mid x+y,z \mid = \mid x,z \mid^y + \mid y,z \mid + wU_1'(x,y,z), \tag{4}$$

$$\mid x,y+z \mid = \mid x,y \mid^z + \mid x,z \mid + wU_2'(x,y,z). \tag{5}$$

The elements U_1', U_2' are defined in (7.5.2)(3),(4). A strongly quadratic pair

$$G \xrightarrow{\hat{\gamma}} \hat{\Gamma}(G,\theta) \xleftarrow{w} \otimes^4 C \tag{6}$$

is *universal* if for each pair (f,w) above there is a unique homomorphism of G-modules $(f,w)^{\square} : \hat{\Gamma}(G,\theta) \to K$ with $(f,w)^{\square}\hat{\gamma} = f$ and $(f,w)^{\square} = w$.

7.5.4. LEMMA: *A strongly quadratic pair (f,w) satisfies*

$$f(0) = 0 = \mid 0,\ x \mid = \mid x, 0 \mid \ \text{ and } \ \mid x, x \mid = 2f(x).$$

We leave the lengthy proof of the lemma to the reader. The next two propositions describe the important examples of pairs which are strongly quadratic in the sense of (7.5.3).

7.5.5. PROPOSITION: *Let $G = < Z >$ be a free group and let (G,θ_0) be the crossed square group in (7.2.3) where $q: G \to \pi = G^{ab}$ is the abelianization. Then the pair*

$$G \xrightarrow{f} \hat{J}_Q G \xleftarrow{w} \otimes^4 C_1$$

defined by f in (7.4.3) and by $w_2 = d\omega$ in (7.4.1) is (G,θ_0)-strongly quadratic.

Proof: We set

$$(1) \qquad f(x) = v(x \bar{\bar{\otimes}} x), \ \mid x, y \mid = v(x \bar{\bar{\otimes}} y + y \bar{\bar{\otimes}} x).$$

Then the distributivity laws for $\bar{\bar{\otimes}}$ in IV.12.3 of [BCH] yield the following formulas in $(J_Q G)_2$ where we write $v(x \bar{\bar{\otimes}} y) = x \bar{\bar{\otimes}} y$.

$$(2) \qquad (x+y) \bar{\bar{\otimes}} (x+y) = (x+y) \bar{\bar{\otimes}} y + ((x+y) \bar{\bar{\otimes}} x)^y$$

$$\begin{aligned} &= (x \bar{\bar{\otimes}} y)^y + y \bar{\bar{\otimes}} y + (x \bar{\bar{\otimes}} x)^{2y} + (y \bar{\bar{\otimes}} x)^y \\ &\quad + \underbrace{wT(x^y y \otimes \theta_0(y))}_{(3)} + wT(x^y y \otimes \theta_0(x))^y \\ &= y \bar{\bar{\otimes}} y + (x \bar{\bar{\otimes}} y)^y - < y \bar{\bar{\otimes}} y, (x \bar{\bar{\otimes}} y)^y > \\ &\quad + (y \bar{\bar{\otimes}} x)^y + (x \bar{\bar{\otimes}} x)^{2y} + < (x \bar{\bar{\otimes}} x)^{2y}, (y \bar{\bar{\otimes}} x)^y > + (3) \end{aligned}$$

Here the Pfeiffer commutators $< a, b >= w(\{a\} \otimes \{b\})$ are commutators since $d(y \bar{\bar{\otimes}} y) = 0$, $d(x \bar{\bar{\otimes}} x) = 0$. These equations show that by (1) the equation (7.5.3)(1) is satisfied. Next we get the equations in (7.5.3)(4),(5) by (4) and

(5).

$$
\begin{array}{ll}
 & (x+y) \mathbin{\bar{\bar{\otimes}}} z + z \mathbin{\bar{\bar{\otimes}}} (x+y) \\
(4) \quad = & x \mathbin{\bar{\bar{\otimes}}} z^y + y \mathbin{\bar{\bar{\otimes}}} z \\
 & +wT(x^y \otimes y \otimes \theta_0(z)) + z \mathbin{\bar{\bar{\otimes}}} y + z \mathbin{\bar{\bar{\otimes}}} x^y \\
= & x \mathbin{\bar{\bar{\otimes}}} z^y + z \mathbin{\bar{\bar{\otimes}}} x^y + y \mathbin{\bar{\bar{\otimes}}} z + z \mathbin{\bar{\bar{\otimes}}} y \\
 & + < y \mathbin{\bar{\bar{\otimes}}} z + z \mathbin{\bar{\bar{\otimes}}} y, z \mathbin{\bar{\bar{\otimes}}} x^y > +wT(x^y y \otimes \theta_0(z)).
\end{array}
$$

$$
\begin{array}{lll}
x \mathbin{\bar{\bar{\otimes}}} (y+z) + (y+z) \mathbin{\bar{\bar{\otimes}}} x & = & x \mathbin{\bar{\bar{\otimes}}} z + x \mathbin{\bar{\bar{\otimes}}} y^z + y \mathbin{\bar{\bar{\otimes}}} x^z \\
 & & +z \mathbin{\bar{\bar{\otimes}}} x + wT(y^z z \otimes \theta_0(x)) \\
(5) & = & x \mathbin{\bar{\bar{\otimes}}} y^z + y \mathbin{\bar{\bar{\otimes}}} x^z + x \mathbin{\bar{\bar{\otimes}}} z \\
 & & - < x \mathbin{\bar{\bar{\otimes}}} y^z + y \mathbin{\bar{\bar{\otimes}}} x^z, x \mathbin{\bar{\bar{\otimes}}} z > +z \mathbin{\bar{\bar{\otimes}}} x \\
 & & +wT(y^z \otimes \theta_0(x)).
\end{array}
$$

Next we get (7.5.3)(3) since

$$
(6) \qquad -x \mathbin{\bar{\bar{\otimes}}} x - y \mathbin{\bar{\bar{\otimes}}} y + x \mathbin{\bar{\bar{\otimes}}} x + y \mathbin{\bar{\bar{\otimes}}} y =< x \mathbin{\bar{\bar{\otimes}}} x, y \mathbin{\bar{\bar{\otimes}}} y >
$$

is a Pfeiffer commutator, $d(y \mathbin{\bar{\bar{\otimes}}} y) = 0$. It remains to check (7.5.3)(2). For this we consider the equations

$$
\begin{array}{lll}
(-x) \mathbin{\bar{\bar{\otimes}}} (-x) & = & -((-x) \mathbin{\bar{\bar{\otimes}}} x)^{-x} \\
(7) \qquad & = & -[-(x \mathbin{\bar{\bar{\otimes}}} x)^{-x} - w_2 \bar{\bar{\Theta}} (x, -x)(x)]^{-x} \\
 & = & (x \mathbin{\bar{\bar{\otimes}}} x)^{-2x} - w_2 T((h_1 x)^{-x} \otimes h_1(-x) \otimes \theta_0(x))^{-x}
\end{array}
$$

where we use IV.12.3(4) and IV.12.4(4) in [BCH]. Since h_1 is a crossed homomorphism we have $h_1(-x) = -h_1(x)^{-x}$ so that (7) implies (7.5.3)(2). □

The next result is the analogue of (6.3.5).

7.5.6. PROPOSITION: *Let $(G, \theta) = (G, A, h, \theta)$ be a square group and let the pair*

$$
G \xrightarrow{\tilde{\tilde{\Gamma}}} \tilde{\tilde{\Gamma}}(G, \theta) \xleftarrow{w} \otimes^4 A
$$

be given by the universal quadratic pair (7.1.6) with $w = \delta\omega I$. *Here* $I : \otimes^4 A \to \phi(A)$ *is the canonical quotient map. Then* $(\tilde{\tilde{\gamma}}, w)$ *is* (G, θ)*-strongly quadratic if we use the structure of* $\tilde{\tilde{\Gamma}}(G, \theta)$ *as a* G*-module in (7.1.6)(11).*

7.5.7. PROPOSITION: *The pair* $(h_2 f, \delta)$ *in (7.5.1) is a* $(G, \epsilon_* \theta_0)$*-quadratic pair in the sense of (7.1.6).*

Proof: Let $x, z \in G$. Since $R(x, x, z) = 0$, see (7.4.5)(4),(5), we see that

$$s(\{x\} \otimes \{x\} \otimes \{z\}) = h_3 v((x \bar{\bar{\otimes}} x) \bar{\bar{\otimes}} z). \tag{1}$$

Whence we get

$$\begin{aligned} [x, x, z] &= \delta s(\{x\} \otimes \{x\} \otimes \{z\}), \\ &= \delta h_3 v((x \bar{\bar{\otimes}} x) \bar{\otimes} z), \\ &= h_2 d v((x \bar{\bar{\otimes}} x) \bar{\otimes} z), \\ &= h_2((d(x \bar{\bar{\otimes}} x)) \bar{\otimes} z - x \bar{\bar{\otimes}} x^z + x \bar{\bar{\otimes}} x), \\ &= -h_2 f(x)^z + h_2 f(x). \end{aligned} \tag{2}$$

Here we use the boundary formula IV.12.2(9)(iii) in [BCH] and the fact that $d(x \bar{\bar{\otimes}} x) = 0$ in $J_Q(G)$. Formula (2) led us to the definition of the G-action in (7.1.16). We now use proposition (7.5.5) and formula (2) for the proof of (7.5.7). First we get by (2) the formula

$$h_2(| x, y |^z) = h_2 | x, y | - [x, y, z] - [y, x, z] \tag{3}$$

since $h_2 w U'(x, y) = \delta\omega U(x, y)$ and since G acts trivially on $\tilde{\otimes}^3 (G, \theta_0)$, compare (6.3.6)(1). For

$$[x, y] = -h_2 f(y) + h_2 f(x + y) - h_2 f(x) \tag{4}$$

we thus get by (7.5.3)(1) the formula

$$(5) \qquad \begin{cases} [x,y] = h_2 \mid x,y \mid -B(x,y) + \delta\omega U(x,y), \\ B(x,y) = 2[x,x,y] + [x,y,y] + [y,x,y], \end{cases}$$

compare (6.3.6)(2). By (5) we derive (7.1.6)(6),(7) from (7.5.3)(4),(5). Moreover (7.1.6)(5) follows from (7.5.3)(2) and (7.1.6)(4) follows from (7.5.3)(2). Whence it remains to check (7.1.6)(3). For this we consider

$$(6) \qquad [y,x,x] = \delta h_3 v((x \overset{=}{\otimes} x) \overset{=}{\otimes} x) - \delta\omega R(y,x,x)$$

where $\delta h_3 = h_2 d$. Equation (6) is a consequence of (7.4.5)(4) and the definition of h_3. By the boundary formulas in IV.12.2(9) of [BCH] we get in $(J_Q G)_2$ the following equations where we write $v(x \overset{=}{\otimes} y) = x \overset{=}{\otimes} y$, $v(x \bar{\otimes} y) = x \bar{\otimes} y$ and where $d(y \overset{=}{\otimes} x) = -y - x + y + x = (x,y)$ is the commutator.

$$(7) \qquad \begin{aligned} dv((x \overset{=}{\otimes} x)\bar{\otimes} x) &= (y,x)\bar{\otimes} x - (y \overset{=}{\otimes} x)^x + y \overset{=}{\otimes} x \\ &= -(y\bar{\otimes} x)^{(y,x)} - (x\bar{\otimes} x)^{y+(y,x)} + (y\bar{\otimes} x)^x \\ &\quad + x\bar{\otimes} x - (y \overset{=}{\otimes} x)^x + y \overset{=}{\otimes} x. \end{aligned}$$

Here we have for $a \in (JQ)_2$ the Pfeiffer commutator formula

$$(8) \qquad a^{(y,x)} = a^{d(y \overset{=}{\otimes} x)} = -y \overset{=}{\otimes} x + a + y \overset{=}{\otimes} x + < y \overset{=}{\otimes} x, a >$$

with $< y \overset{=}{\otimes} x, a > = w(yx \otimes \{a\})$. Moreover we have the formula $(\theta = \theta_0)$

$$(9) \qquad x\bar{\otimes} y = x \overset{=}{\otimes} y - wT(\theta x \otimes \theta y),$$

compare IV.12.4(7) [BCH]. Using (6)...(9) we get in $\overset{\approx}{J}G$ the following formulas where we omit δw and h_2.

$$
\begin{array}{rcl}
[y,x,x] & = & dv((y \mathbin{\overline{\overline{\otimes}}} x)\bar{\otimes} x) - R(y,x,x) \\
 & = & -\, y \mathbin{\overline{\overline{\otimes}}} x - y \mathbin{\overline{\overline{\otimes}}} x + y \mathbin{\overline{\overline{\otimes}}} x - T(\theta y \otimes \theta x) + yx\hat{\otimes} yx \\
 & & -\, y \mathbin{\overline{\overline{\otimes}}} x - (x \mathbin{\overline{\overline{\otimes}}} x)^y + y \mathbin{\overline{\overline{\otimes}}} x - T(\theta x \otimes \theta x) + yx\hat{\otimes} xx \\
 & & +\, (y \mathbin{\overline{\overline{\otimes}}} x)^x + T(\theta y \otimes \theta x) \\
(10) & & +\, x \mathbin{\overline{\overline{\otimes}}} x + T(\theta x \otimes \theta x) \\
 & & -\, (y \mathbin{\overline{\overline{\otimes}}} x)^x + y \mathbin{\overline{\overline{\otimes}}} x - R(y,x,x), \\
 & = & -\, (x \mathbin{\overline{\overline{\otimes}}} x)^y + x \mathbin{\overline{\overline{\otimes}}} x + yx\hat{\otimes} yx + yx\hat{\otimes} xx - R(y,x,x) \\
 & = & [x,x,y] + \nabla(x,y).
\end{array}
$$

□

Proposition (7.5.6) implies that we have for the crossed square group (G, θ_0) in (7.2.3) the commutative diagram

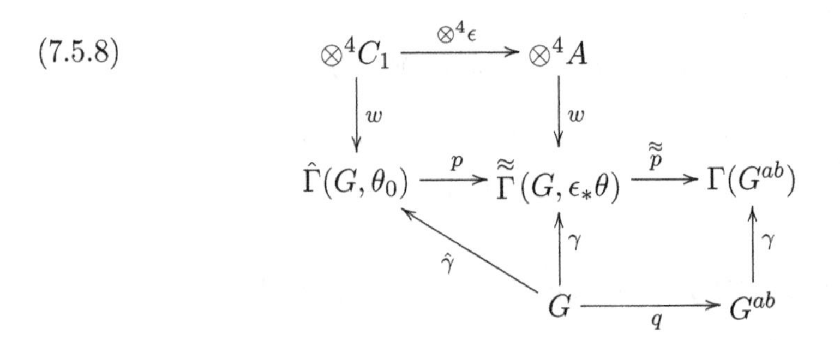

where $p = (\gamma, w(\otimes^4 \epsilon))^{\square}$ and where we use the augmentation map ϵ in (7.2.3)(4). Diagram (7.5.8) is the analogue of diagram (6.3.8). The maps p and $\approx\!\!\!\!p$ in (7.5.8) are G-equivariant. We shall use the following property of the composition $\hat{p} = \approx\!\!\!\!p\, p$ which corresponds to (6.3.9).

7.5.9. PROPOSITION: *Let (G, θ_0) be the crossed square group in (7.2.3). Then the elements $\hat{\gamma} x - (\hat{\gamma} x)^{\alpha}$, $x \in G$, $\alpha \in G$, and the elements in the image of w generate the kernel of $\hat{p} = \approx\!\!\!\!p\, p$ in (7.5.8) as a G-module.*

The proof of (7.5.9) uses almost literally the same arguments as the proof of (6.3.9). We are now ready for the proof of theorem (7.5.1). For this we follow the lines of the proof of (6.3.1) in (6.3.10).

7.5.10. Proof (of (7.5.1)): We obtain by (7.5.5) the commutative diagram

$$(1)\qquad \begin{array}{ccc} \hat\Gamma(G,\theta_0) & \xrightarrow{p} & \overset{\approx}{\Gamma}(G,\epsilon_*\theta_0) \\ {\scriptstyle f^{\square}=(f,w)^{\square}}\downarrow & & \downarrow{\scriptstyle (h_2 f,\delta)^{\square}=k} \\ \hat J_Q G & \xrightarrow[h_2]{} & \overset{\approx}{J} G \end{array}$$

where p is the map in (7.5.8) and where k is defined by the universal property. The map h_2 is constructed in (7.4.6). The composition $kp = h_2 f^{\square}$ of maps in (1) is used in the following commutative diagram where we construct an inverse k' of k in (1).

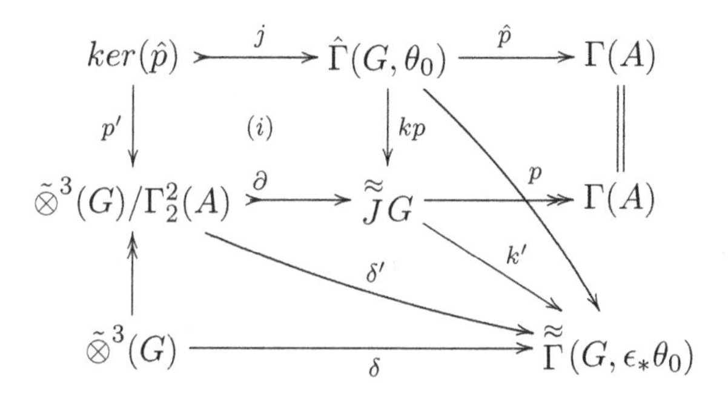

Here the rows are short exact sequences so that (i) is a central push out. The map ∂ is induced by δ in (7.4.9) and δ' is induced by δ since $\delta i = 0$, see (7.1.15). As in (6.3.10) we see by (7.5.9) that $\delta' p' = pj$ holds so that there is a unique homomorphism k' which extends the diagram commutatively. As in (6.3.10) this implies that k' is the inverse of k and that k is an isomorphism. It remains to check that k is natural, this follows by the next proposition. □

7.5.11. PROPOSITION: *The isomorphism k is natural for homomorphisms $\eta : G' \to G$ between free groups. Here we use the generalized induced maps in (7.1.7).*

Proof: The induced map η_* for $\overset{\approx}{J}G$ is given by the commutative diagram

$$\begin{array}{ccccc} (G'\otimes G')_2 & \xrightarrow{\nu} & (J_Q G')_2 & \xrightarrow{h_2} & (\overset{\approx}{J}G')_2 \\ \downarrow{\scriptstyle \eta\otimes\eta} & & \downarrow{\scriptstyle \eta_2} & & \downarrow{\scriptstyle \eta_*} \\ (G\otimes G)_2 & \xrightarrow[\nu]{} & (J_Q G)_2 & \xrightarrow[h_2]{} & (\overset{\approx}{J}G)_2 \end{array}$$

We obtain η_* as in (7.4.6)(5). We now compute for $x, y \in G$ the difference

$$A(x,y) = h_2(\eta_2 v(x \mathbin{\bar{\bar{\otimes}}} y) - v(\eta x \mathbin{\bar{\bar{\otimes}}} \eta y)). \tag{1}$$

By definition of $\eta \otimes \eta$ in IV.12.10 of [BCH] and (7.3.4)(1) we see that $A(e,f) = 0$ for generators $e, f \in Z'$ in G'.

Moreover we obtain by IV.12.4 of [BCH] the following equations where we omit h_2 and v in the notation:

$$\begin{aligned} A(x+x',y) &= \eta_2((x+x') \mathbin{\bar{\bar{\otimes}}} y) - \eta(x+x') \mathbin{\bar{\bar{\otimes}}} \eta y \\ &= \eta_2[(x \mathbin{\bar{\bar{\otimes}}} y)^{x'} + x' \mathbin{\bar{\bar{\otimes}}} y] + \eta_* T(xx' \otimes \theta(y)) \\ &\quad - [(\eta x \mathbin{\bar{\bar{\otimes}}} \eta y)^{\eta x'} + \eta x' \mathbin{\bar{\bar{\otimes}}} \eta y] - T(\eta_*(xx') \otimes \theta(\eta y)) \\ &= A(x,y) + A(x',y) + T(\eta_*(xx') \otimes \theta(\eta)(y)) \end{aligned} \tag{2}$$

$$A(x,y) = \eta_2(x \bar{\otimes} y) - \eta x \bar{\otimes} \eta y + T(\theta x \otimes \theta y) - T(\theta \eta x \otimes \theta \eta y). \tag{3}$$

$$\begin{aligned} A(x, y+y') &= \eta_2[x \bar{\otimes} y' + (x \bar{\otimes} y)^{y'}] - \eta_* T(\theta x \otimes yy') \\ &\quad - [\eta x \bar{\otimes} \eta y' + (\eta x \bar{\otimes} \eta y)^{\eta y'}] + T(\theta \eta x \otimes \eta_*(yy')) \\ &\quad + \eta_* T(\theta x \otimes \theta(y+y')) - T(\theta \eta x \otimes \theta \eta(y+y')) \\ &= A(x,y) + A(x,y'). \end{aligned} \tag{4}$$

Now (4) implies that $A(e,y) = 0$ for all $y \in G'$, $e \in Z'$ since $A(e,f) = 0$ for $e, f \in Z'$. This shows that $A(x,y) = w\nabla_\eta$ where ∇_η is the map in (7.1.7)(1). In fact, the map $w\nabla_\eta$ satisfies the same law as in (2) and (4) and $w\nabla_\eta(e,f) = 0$ for $e, f \in Z'$. The equation $A = w\nabla_\eta$ implies that the isomorphism k is natural for the generalized induced maps in (7.1.7). □

7.6. The 3-type of the loop space of a one point union of 2-spheres

We compute a quadratic module which characterizes the 3-type of the loop space $\Omega M(A,2)$ where $M(A,2)$ is a one point union of 2-spheres.

Let σ be a totally free quadratic chain complex. We say that a quadratic chain complex σ' is the 3-*type* of σ if $(\sigma')_j = 0$ for $j > 3$ (so that σ' is the same as a quadratic module, see [BCH]) and if there is a map $h : \sigma \to \sigma'$ which induces isomorphisms $\pi_j h : \pi_j \sigma \cong \pi_j \sigma'$ for $j \le 3$. We know by IV.§10 of [BCH] that a quadratic module is the *algebraic equivalent* of a topological 3-type. We now describe the quadratic module which characterizes the 3-type of the loop space $\Omega M(A,2) \simeq JM_A$ where A is a free abelian group. This quadratic module can be obtained by the 3-type of the quadratic chain complex $J_Q(G)$ with $G = \pi_1 M_A$ since we have the isomorphism $J_Q(G) \cong \sigma(JM_A)$, see (7.3.6). The next result, in (7.6.4), is the quadratic analogue of theorem (6.4.1) where we describe the 3-type of $J(G) = \lambda J_Q(G)$.

For this first recall that we have by (7.1.17) the following natural commutative diagram of exact sequences where G is a free group with $G^{ab} = A$.

(7.6.1)
$$\begin{array}{ccccccc}
\Gamma_2^2 A & \rightarrowtail & \tilde{\otimes}^3(G,\theta_0) & \xrightarrow{\theta} & \tilde{\tilde{\Gamma}}(G,\theta_0) & \xrightarrow{\tilde{\tilde{p}}} & \Gamma A \\
\| & & \downarrow p_3 & & \downarrow h & & \| \\
\Gamma_2^2 & \overset{i}{\rightarrowtail} & \otimes_2^2 A & \xrightarrow{\delta} & \bar{\tilde{\Gamma}}(G,q\theta_0) & \xrightarrow{\bar{\tilde{p}}} & \Gamma A
\end{array}$$

Here we have $\otimes_2^2 A = \otimes^3 A \oplus \tilde{\otimes}^4 A$ and h and p_3 are given by (h,h) and (p, Ip) respectively, see (7.1.17) and (7.1.10)(2). The bottom row of (7.6.1) is obtained by the pull back in (7.1.10)(2). We deduce from (7.6.1) and (7.5.1) the central push out diagram

(7.6.2)
$$\begin{array}{ccccc}
\tilde{\otimes}^3(G,\theta_0) & \xrightarrow{\delta} & \tilde{\tilde{\Gamma}}(G,\theta_0) & \overset{j}{\rightarrowtail} & (\tilde{\tilde{J}}G)_2 \\
\downarrow p_3 & c\ push & \downarrow h & c\ push & \downarrow p_2 \\
\otimes_2^2 A & \xrightarrow{\delta} & \bar{\tilde{\Gamma}}(G,q\theta_0) & \overset{j}{\rightarrowtail} & (\bar{\tilde{J}}\,G)_2
\end{array}$$

which defines the group $(\bar{\tilde{J}}\,G)_2$. The map j in the top row is given by the isomorphism in (7.5.1). We now derive from (7.6.2) the following diagram of C-groups which determines the quadratic module $\chi(G)$ below.

(7.6.3)
$$\begin{array}{ccccc}
\otimes^4 A & & & & \\
\downarrow \omega & & & & \\
\otimes_2^2 A & \xrightarrow{j\delta} & (\bar{\tilde{J}}G)_2 & \xrightarrow{\bar{\tilde{d}}} & G
\end{array}$$

Her $\bar{\tilde{d}}$ is induced by $\tilde{\tilde{d}}$ in (7.4.6)(1) and ω is given by $I : \otimes^4 A \to \tilde{\otimes}^4 A$ and by the inclusion $\tilde{\otimes}^4 A \subset \otimes_2^2 A$. Using j in (7.6.2) we obtain the natural identification

(1)
$$\overset{\approx}{\Gamma}(G, q\theta_0) = kernel(\overset{\approx}{d}).$$

The action of G on $(\overset{\approx}{J}G)_2$ is given by the action of G on $(\tilde{\overline{J}}G)_2$ and by the trivial action on $\otimes_2^2 A$. The action of G on $\overset{\approx}{\Gamma}$ coincides via (7.5.1) with the G^{ab}-action in (7.1.6)(11). Moreover G acts trivially on $\otimes_2^2 A$. We obtain by ω in (7.6.3) the following quadratic map ω':

(2)
$$\omega' : C' \otimes C' \xrightarrow{q' \otimes q'} \otimes^4 A \xrightarrow{\omega} \otimes_2^2 A.$$

Here C' is the abelianization of the cokernel of $\omega = j\delta\omega$. Using the map qp in (7.4.1) we obtain the induced map q' by the commutative diagram, see (7.4.3),

(3)
$$\begin{array}{ccccc} (\overset{\approx}{J}G)_2 & \twoheadrightarrow & C' & \xrightarrow{q'} & \otimes^2 A \\ \uparrow{\scriptstyle p_2h_2} & & & & \| \\ (J_QG)_2 & & \xrightarrow{qp} & & \otimes^2 G^{ab} \end{array}$$

Now one can check that $(\omega', j\delta, \overset{\approx}{d}) = \chi(G)$ satisfies the properties of a quadratic module in the sense of IV.1.10 in [BCH]. Moreover we get:

7.6.4. THEOREM: *Let G be a free group with $A = G^{ab}$. Then the 3-type of the loop space $\Omega M(A, 2)$ is naturally given by the quadratic module $\chi(G)$ in (7.6.3).*

Proof: We know that the 3-type of the quadratic chain complex J_QG describes naturally the 3-type of $\Omega M(A, 2)$. We now obtain a natural 3-equivalence

(1)
$$h : J_QG \longrightarrow \chi(G)$$

by use of the map $h = (p_3h_3, p_2h_2, h_1 = 1_G)$ where we use h_3, h_2, h_1 in (7.4.3) and the maps p_3, p_2 in (7.6.2). Using diagram (7.6.1) we see that h is actually a 3-equivalence. □

7.6.5. REMARK: Recall that we have the functor λ which carries quadratic chain complexes to crossed chain complexes. The functor λ is simply given by dividing out the image of w and ω, see IV.3.2 in[BCH]. If we apply λ to the quadratic module in (7.6.3) we get the 3-type of $J(G) = \lambda J_Q(G)$ described in (6.4.1). In particular C' in (7.6.3) is the abelianization of $(\tilde{J}G)_2$.

7.7. Algebraic models of tracks in $\mathbf{T}(2, 4)$

We apply the quadratic chain complex functor σ to sets of topological tracks. Using the quadratic module in (7.6) this yields algebraic models of tracks which we use for the computation of the topological track category $\mathbf{T}(2, 4)$, in particular we obtain a proof of theorem (5.2.4).

For the pull back $\mathbf{T}$ in (5.2.1) we obtain the equivalence of categories

(7.7.1) $$\chi : \mathbf{T}(2, 4) \to \mathbf{T}$$

as follows. We proceed the same way as in (6.4.3) and (6.4.4). We simply replace the functor ρ in (6.4.3)(2) by the functor σ and we replace the crossed 3-type of JG by the 3-type in (7.6.3). As in (6.4.3)(6) we obtain $\tilde{\bar{g}}$ and the commutative diagram;

(1) $$\begin{array}{ccc} E_B = (\sigma M(B, 2))_2 & \xrightarrow{\bar{g}_*} & (\sigma J M_A)_2 = (J_Q G)_2 \\ \downarrow{\scriptstyle \tilde{\bar{g}}} & & \downarrow{\scriptstyle h_2} \\ \tilde{\bar{\Gamma}}(E_A, \theta_0) = \tilde{\bar{\Gamma}}(G, \theta_0) = ker(\tilde{\bar{d}}) & \rightarrowtail & (\tilde{\bar{J}}\, G)_2 \end{array}$$

Here M_A is a one point union of 1-spheres with $\pi_1 M_A = G$, $A = G^{ab}$ and $M(B, 2)$ is a one point union of 2-spheres. By IV Appendix C in [BCH] we see that the free nil(2)-group E_B determines the quadratic chain complex $\sigma M(B, 2)$.

The method in (6.4) now yields the following bijection with $Y = M_A$, $X = M(B', 2)$ as in (6.4.3).

(2) $$\chi : T(g(\Sigma x), (\Sigma y)f) = [I_* X, JY]^{\bar{g}x, J(y)\bar{f}}$$

$$
\begin{array}{rcl}
(3) & \cong & [\sigma I_* X, \sigma JY]^{(\bar{g}x)_*,(J(y)\bar{f})_*} \\
(4) & \cong & [I_* E_{B'}, J_Q G]^{\bar{g}_* x_*, J(y_*)\bar{f}_*} \\
(5) & \cong & [I_* E_{B'}, \chi(G)]^{h\bar{g}_* x_*, hJ(y_*)\bar{f}_*} \\
(6) & = & [I_* E_{B'}, \delta]^{\tilde{\bar{g}}\xi, \tilde{\bar{\Gamma}}(\eta)\tilde{\bar{f}}}.
\end{array}
$$

Here (3) is obtained by the functor σ in (7.3.6), for this we use the theorem IV.8.2 of[BCH]. Now (5) is induced by the 3-equivalence h in (7.6.4)(1). This shows by definition of the set in (5) that (6) is satisfied. Here (6) denotes the set of tracks H for which (ξ, η, H) is a morphism $\tilde{\bar{g}} \to \tilde{\bar{f}}$ in $\mathbf{T}$. The bijection (2)...(6) yields the equivalence in (7.7.1) above. The equivalence of pseudo functors in (7.3.6) shows that χ is a well defined functor, compare (5.1.19) and (7.3.5)(4), (7.3.5)(5). This completes the proof of (5.2.3).

We now consider the following diagram:

(7.7.2)
$$
\begin{array}{ccc}
 & \Sigma x' \text{ (upper arc)}, \ \Uparrow H_\xi & \\
M(B',2) & \xrightarrow{\quad \Sigma x \quad} & M(B,2) \\
\Big\downarrow \bar{f} & \overset{H}{\Longrightarrow} & \Big\downarrow \bar{g} \\
JM_{A'} & \xrightarrow{\quad J(y) \quad} & JM_A \\
 & \Downarrow J(H_\eta), \ J(y') \text{ (lower arc)} &
\end{array}
$$

The pasting of these tracks yields the sum of tracks in (3.2.1)(3). The homotopy H_ξ corresponds (via the functor σ) to the homotopy

(1) $$H_\xi : \xi \cong \xi', \ \xi, \xi' : \sigma(M(B', 2)) = E_{B'} \to E_B = \sigma(M(B, 2))$$

of maps in $\mathbf{Q}$, that is, H_ξ is a homomorphism

(2) $$H_\xi : E_{B'} \to B' \to \otimes^2 B = \sigma(M(B, 2))_3,$$

this is a 0-homotopy in the sense of IV.C.7(3) in [BCH]. Now $\bar{g}$ in (7.7.2) induces the following map $g_* = (g_1, g_2, g_3)$ between quadratic modules where the bottom row is $\chi(G)$ in (7.6.3).

$$\begin{array}{ccccccc}
\otimes^2 B & \xrightarrow{\omega} & \otimes^2 B & \xrightarrow{d} & E_B & \xrightarrow{d} & 0 \\
\downarrow{\scriptstyle \otimes^2 \tau g} & & \downarrow{\scriptstyle \Gamma_3} & & \downarrow{\scriptstyle g_2} & & \downarrow{\scriptstyle g_1} \\
A^{\otimes 4} & \xrightarrow{\omega} & \otimes_2^2 A & \xrightarrow{\delta} & (\tilde{\bar{J}}\, G)_2 & \xrightarrow{d} & G
\end{array}$$

Here ω in the top row is the identity; we see that g_3 is given by $\omega(\otimes^2 \tau g)$ with

$$g_* = \tau g : B \xrightarrow{g} \Gamma A \longrightarrow \otimes^2 A, \;\; g_* = C_2(g).$$

This shows that the identification χ in (7.7.1) satisfies the following addition formula:

(7.7.3)
$$\begin{cases} \chi(H + \bar{g}_* H_\xi) & = & \chi(H) + g_\# H_\xi \\ \text{with } g_\# H_\xi & = & \omega(\otimes^2 \tau g) H_\xi \end{cases}$$

Here the right hand side is addition of homomorphisms of the form $B' \to \otimes_2^2 A$, see (5.1.22)(2), and the left hand side is given by pasting tracks in (7.7.2).

Next we consider an addition formula for $\chi(-\bar{f}^* J(H_\eta) + H)$ in (7.7.2). This formula is more complicated, we proceed for this in the same way as in (6.4.5) where we replace the functor ρ of crossed chain complexes by the functor σ of quadratic chain complexes. This way we get the formula

(7.7.4)
$$\chi(-\bar{f}^* J(H_\eta) + H) = -\bar{H}_\eta f + \chi(H)$$

where we use the definiton of $\bar{H}_\eta$ in (5.1.18)(3), compare (5.1.18)(4).

Finally we obtain a proof of (5.2.4) as follows. We use (7.7.1), (7.7.3) and (7.7.4). Moreover we replace the pull back $\mathbf{T}$ in (5.2.1) by the split extension (5.2.2). For this we need the section (4.5.3).

CHAPTER 8

ON THE COHOMOLOGY OF THE CATEGORY NIL.(T. PIRASHVILI)

Appendix by T. Pirashvili

8.1. Introduction

Let **nil** be the category of finitely generated free groups of nilpotency degre[e] two. For each abelian group M one can consider the bifunctor

$$\mathcal{D}_M : \mathbf{nil}^{op} \times \mathbf{nil} \longrightarrow \mathbf{Ab}$$

given by

$$\mathcal{D}_M(X,Y) = Hom(\Gamma X_{ab}, L(Y_{ab},1)_3 \otimes M).$$

The aim of this appendix is to prove the following.

8.1.1. THEOREM: *The low dimensional cohomology of* **nil** *with coefficients i[n]* $\mathcal{D}_M$ *is given by*

$$H^i(\mathbf{nil},\mathcal{D}_M) = \begin{cases} 0 & , \quad i=0, \\ M & , \quad i=1, \\ {}_2M & , \quad i=2. \end{cases}$$

Here ${}_2M = \{m \mid 2m = 0\}$.

The idea of the proof is to pass to the cohomology of the category **ab** of [fi]nitely generated free abelian groups, which are known as MacLane cohomolog[y] [JP] or topological Hochschild cohomology and use the vanishing result of t[he] author (see the appendix of [BeP]).

8.2. General facts

Let $\mathbf{C}$ be a small category and let $\mathcal{F}(\mathbf{C})$ be the category of all functors $\mathbf{C} \to \mathbf{Ab}$. For $C \in \mathcal{O}b\,\mathbf{C}$ we denote by h_C the functor given by

$$X \mapsto \mathbb{Z}[\mathbf{C}(C, X)], \;\; X \in \mathcal{O}b\,\mathbf{C}.$$

Then $h_C \in \mathcal{O}b\,\mathcal{F}(\mathbf{C})$ and for any $T \in \mathcal{F}(\mathbf{C})$ one has

$$Hom_{\mathcal{F}(\mathbf{C})}(h_C, T) \cong T(C).$$

Hence h_C is a projective object in $\mathcal{F}(\mathbf{C})$ and the collection $(h_C)_{C \in \mathcal{O}b\,\mathbf{C}}$ is a set of small projective generators of $\mathcal{F}(\mathbf{C})$. In particular $\mathcal{F}(\mathbf{C})$ is an abelian category with enough projective and injective objects. Let $U, T : \mathbf{C} \to \mathbf{Ab}$ be functors. Then one can define the bifunctor $\mathcal{D} : \mathbf{C}^{op} \times \mathbf{C} \to \mathbf{Ab}$ by

$$\mathcal{D}(X, Y) = Hom(U(X), T(Y)).$$

By corollary 3.11 of [JP] one has an isomorphism

$$H^*(\mathbf{C}, \mathcal{D}) \cong Ext^*_{\mathcal{F}(\mathbf{C})}(U, T), \tag{8.2.1}$$

provided $U(X)$ is a free abelian group for all $X \in \mathcal{O}b\mathbf{C}$.

Let $\mathbf{D}$ also be a small category and let $F : \mathbf{C} \to \mathbf{D}$ be a functor. Then the precomposition with F induces a functor

$$\begin{array}{rlcl} F^* : & \mathcal{F}(\mathbf{D}) & \longrightarrow & \mathcal{F}(\mathbf{C}), \\ & \mathcal{F}(\mathbf{D}) \ni T & \mapsto & T \circ F. \end{array}$$

It is well known that F^* has the left adjoint functor, known as left Kan extension (see [Mac]),

$$F_! : \mathcal{F}(\mathbf{C}) \longrightarrow \mathcal{F}(\mathbf{D}).$$

Thus $F_!$ is a right exact functor and

$$F_!(h_C) = h_{F(C)}.$$

We let $L_*F_!$ be the left derived functors of $F_!$. Then the Grothendieck spectral sequence for the composite functors yields the following spectral sequence:

$$E_2^{pq} = Ext^p_{\mathcal{F}(\mathbf{D})}((L_qF_!)U, T) \Rightarrow Ext^{p+q}_{\mathcal{F}(\mathbf{C})}(U, T \circ F). \tag{8.2.2}$$

Here $U : \mathbf{C} \to \mathbf{Ab}$ and $T : \mathbf{D} \to \mathbf{Ab}$ are arbitrary functors.

8.3. $L_*F_!$ and simplicial derived functors

In this section we consider the abelianization functor $G \to G/[G,G]$. It take free nilpotent groups to free abelian groups and hence yields a functor

$$F : \mathbf{nil} \longrightarrow \mathbf{ab}.$$

By the previous section we have the functors

$$L_nF_! : \mathcal{F}(\mathbf{nil}) \longrightarrow \mathcal{F}(\mathbf{ab}),\ n \geq 0.$$

The following lemma is a variant of a result of André and Ulmer (see [Ul] and it relates the values of $L_*F_!$ on $U \in \mathcal{F}(\mathbf{nil})$ to simplicial derived functor of U. For the purposes of this appendix we define the last groups in a rathe restricted form. Let $A \in \mathbf{ab}$ be a finitely generated free abelian group. Reca that a nilpotent group of class 2 is finitely generated if and only if the abel ization is finitely generated. Based on this fact it is easy to show that ther exists a simplicial object $K(A,0)^{Nil}$ in **nil** such that

$$\pi_i(K(A,0)^{Nil}) = \begin{cases} 0 & \text{if } \ i > 0, \\ A & \text{if } \ i = 0. \end{cases}$$

Then the homotopy groups of $U(K(A,0)^{Nil})$ are independent from the choic of $K(A,0)^{Nil}$. Therefore the formula

$$L_*^sU(A) := \pi_*U(K(A,0)^{Nil})$$

gives the well defined functors $L_n^sU \in \mathcal{F}(\mathbf{ab}), n \geq 0$.

8.3.1. LEMMA: *For any $U \in \mathcal{F}(\mathbf{nil})$ one has a natural isomorphism in $\mathcal{F}(\mathbf{ab}$*

$$(L_*F_!)(U) \cong L_*^sU.$$

Proof: Since $F_!$ is a right exact functor between abelian categories with enoug projective objects, the left derived functors of $F_!$ have the following propertie

(i) For any short exact sequence

$$0 \to U_1 \to U \to U_2 \to 0$$

in $\mathcal{F}(\mathbf{nil})$ there is a natural long exact sequence in $\mathcal{F}(\mathbf{ab})$

$$\cdots \to (L_{n+1}F_!)(U_2) \to (L_n F_!)(U_1) \to (L_n F_!)(U) \to \cdots .$$

(ii) $L_n F_!(\bigoplus U_i) \cong \oplus L_n F_!(U_i)$

(iii) For any $C \in \mathbf{nil}$, one has isomorphisms

$$L_n F_! h_C \equiv \begin{cases} 0 & \text{if } n > 0, \\ h_{F(\mathbf{C})} & \text{if } n = 0. \end{cases}$$

Moreover these properties characterize $L_* F_!$ completely. Therefore we need to show that similar properties hold for $U \mapsto L_*^s U$. The first two properties follow from the fact that homotopy groups of simplicial abelian groups commute with direct sums and give a long exact sequence associated to a short exact sequence of simplicial abelian groups. Now take $C \in \mathbf{nil}$. Then $h_C(X) = \mathbb{Z}[X^n], X \in \mathbf{nil}$, where n is a number of free generators in C. Similarly $h_{FC}(A) = \mathbb{Z}[A^n], A \in \mathbf{ab}$. By definition one has

$$\begin{aligned} L_*^s h_C(A) &= \pi_* h_C(K(A,0)^{Nil}) \\ &= H_*((K(A,0)^{Nil})^n). \end{aligned}$$

Since $K(A,0)^{Nil}$ is weakly equivalent to the constant simplicial object, we obtain

$$L_i^s h_C(A) = \begin{cases} \mathbb{Z}[A^n] = h_{FC}(A) & \text{if } i = 0, \\ 0 & \text{if } i > 0, \end{cases}$$

and the lemma follows. □

Now we compute $L_i^s U$, $0 \le i \le 2$ for $U = \Gamma \circ F$. Hence $U(G) = \Gamma(G/[G,G])$.

Let us note that the groups

$$L_*^s F(A) = \pi_*(FK(A,0)^{Nil})$$

are the same as $H^{Nil}_{*+1}(A)$ of [BaP]. According to the ob. if A is a free abelia group, then one has the following natural isomorphisms:

$$\begin{aligned} L^s_0F(A) &= A, \\ L^s_1F(A) &= \Lambda^2 A, \\ L^s_2F(A) &= A \otimes \Lambda^2 A. \end{aligned} \tag{8.3.2}$$

By definition one has

$$L^s_*U(A) = \pi_*(\Gamma X_*),$$

where $U = \Gamma \circ F$, and $X_* = F(K(A,0)^{Nil})$. The formula (8.3.2) gives the val ues of $\pi_i X_* = L^s_i F(A)$ for $i = 0, 1, 2$. Now one can use the universal coefficien sequence for the functor Γ (see [BaP], Theorem 3.2) to obtain in the notatio of [BaP] an isomorphism

$$\pi_n(X_* \otimes \mathbb{Z}^\Gamma) \equiv (\pi_n(X_*) \otimes \mathbb{Z}^\Gamma)_n.$$

for $n = 0, 1, 2$. The last groups can now easily be calculated using section 4 c [BaP]. Finally we obtain

8.3.3. LEMMA: *One has the following natural isomorphisms:*

$$\begin{aligned} L^s_0U(A) &= \Gamma A, \\ L^s_1U(A) &= (A \otimes \Lambda^2 A) \oplus (\Lambda^2 A \otimes \mathbb{Z}/2), \\ L^s_2U(A) &= (A \otimes \Lambda^2 A \otimes A) \oplus (A \otimes \Lambda^2 A \otimes \mathbb{Z}/2) \oplus \Lambda^2(\Lambda^2 A). \end{aligned}$$

8.4. Proof of the theorem

By (8.2.1) one has an isomorphism

$$H^*(\mathbf{nil}, \mathcal{D}_M) \cong Ext^*_{\mathcal{F}(\mathbf{nil})}(U, T \circ F)$$

where $U = \Gamma \circ F$ and $T = L(\_, 1)_3 \otimes M$. So one can apply the spectral se quence (8.2.2). According to lemma (8.3.1) it has the following form:

$$E_2^{pq} = Ext^p_{\mathcal{F}}(L^s_q U, T) \Rightarrow H^{p+q}(\mathbf{nil}, \mathcal{D}_M). \tag{8.4.1}$$

Here $\mathcal{F} = \mathcal{F}(\mathbf{ab})$. By lemma (8.3.3) one has

$$\begin{aligned} E_2^{p0} &= Ext^p_{\mathcal{F}}(\Gamma, T), \\ E_2^{p1} &= Ext^p_{\mathcal{F}}(Id \otimes \Lambda^2, T) \oplus Ext^p_{\mathcal{F}}(\Lambda^2 \otimes \mathbb{Z}/2, T), \\ E_2^{p2} &= Ext^p_{\mathcal{F}}(\otimes^2 \otimes \Lambda^2, T) \oplus Ext^p_{\mathcal{F}}(Id \otimes \Lambda^2 \otimes \mathbb{Z}/2, T) \oplus Ext^p_{\mathcal{F}}(\Lambda^2 \circ \Lambda^2, T). \end{aligned}$$

Here $Id \in \mathcal{F}$ is the inclusion functor $\mathbf{ad} \subset \mathbf{Ab}$. Now lemma (8.5.5) below shows that $E_2^{*,0} = 0$. The second summand of E_2^{*1} vanishes completely thanks to lemma (8.5.6). However, the first summand is nontrivial in general and according to lemma (8.5.7) we have

$$E_2^{11} = 0, \ E_2^{0,1} = M.$$

Furthermore the first and the third summand of $E_2^{02,}$ vanish by lemma (8.5.2). The short exact sequence

$$0 \to id \otimes A \xrightarrow{2} id \otimes \Lambda^2 \to id \otimes \Lambda^2 \otimes \mathbb{Z}/2 \to 0$$

together with lemma (8.5.7) below yields an isomorphism

$$Hom_{\mathcal{F}}(id \otimes \Lambda^2 \otimes \mathbb{Z}/2, T) \equiv {}_2M.$$

This computes $E_2^{0,2} = {}_2M$ and the theorem follows.

8.5. Calculations of $Ext^*_{\mathcal{F}}$.

The starting point is the following vanishing result of the author (see for example the appendix of ([BeP]).

8.5.1. LEMMA: *For any integer n and any abelian group A one has*

$$Ext^*_{\mathcal{F}}(A \otimes id^{\otimes n}, V) = 0 = Ext^*_{\mathcal{F}}(V, A \otimes id^{\otimes n}).$$

Here $V : \mathbf{ab} \to \mathbf{Ab}$ is a functor of degree $< n$.

Let us recall that T denotes the functor

$$L(\_, 1)_3 \otimes M$$

It is clear that $deg(T) = 3$.

8.5.2. LEMMA:

$$Hom_{\mathcal{F}}(\Lambda^2 \circ \Lambda^2, T) = 0 = Hom_{\mathcal{F}}(id^{\otimes 2} \otimes \Lambda^2, T).$$

Proof: By lemma (8.5.1) we have $Ext^*_{\mathcal{F}}(Id^{\otimes 4}, T) = 0$. In particular $Hom_{\mathcal{F}}(Id^{\otimes 4}, T) = 0$ and the lemma follows from the fact that there exis natural epimorphisms $Id^{\otimes 4} \to \Lambda^2 \circ \Lambda^2$ and $Id^{\otimes 4} \to Id^{\otimes 2} \otimes \Lambda^2$. □

In order to do further calculations we need the following definition.

For $S, T \in \mathcal{F}$ let

$$\underline{Ext}^q(S, R) : \mathbf{ab} \to \mathbf{Ab}, \ q \geq 0,$$

be the functor given by

$$\underline{Ext}^q(S, R)(A) = Ext^q_{\mathcal{F}}(S, R(A \otimes \_)).$$

Moreover there exists a spectral sequence

$$E_2^{pq} = Ext^p_{\mathcal{F}}(S, \underline{Ext}^q(R, Q)) \Rightarrow Ext^{p+q}_{\mathcal{F}}(S \otimes R, Q) \tag{8.5.3}$$

(see for example proposition 1.4 of [Be]). Here S, R, T are objects from $\mathcal{F}$.

8.5.4. LEMMA:

$$Ext^*_{\mathcal{F}}(Id^{\otimes 2}, T) = 0$$

Proof: Since

$$L(A \oplus B, 1)_3 = L(A, 1)_3 \oplus L(B, 1)_3 \oplus A^{\otimes 2} \otimes B \oplus A \otimes B^{\otimes 2}$$

one has

$$T(A \oplus \_) = T(A) \oplus T \oplus A^{\otimes 2} \otimes M \otimes Id \oplus A \otimes M \otimes Id^{\otimes 2}.$$

Therefore

$$\underline{Ext}^*(Id, T)(A) = Ext^*_{\mathcal{F}}(Id, T) \oplus Ext^*_{\mathcal{F}}(Id, M \otimes Id) \otimes A^{\otimes 2},$$

because $Ext^*_{\mathcal{F}}(Id, B) = 0 = Ext^*_{\mathcal{F}}(I, B \otimes Id^{\otimes 2})$ thanks to lemma (8.5.1). Here B is any abelian group. Hence we have

$$\underline{Ext}^*(Id, T) = B_1^* \oplus B_2^* \otimes Id^{\otimes 2},$$

where $B_1^* = Ext^*_{\mathcal{F}}(Id, T)$ and $B_2^* = Ext^*_{\mathcal{F}}(Id, M \otimes Id)$. For the same reasons

$$Ext^*_{\mathcal{F}}(I, \underline{Ext}(Id, T)) = 0$$

and the result follows from the spectral sequence (8.5.3). □

8.5.5. LEMMA:

$$Ext^*_{\mathcal{F}}(\Gamma, T) = 0 = Ext^*_{\mathcal{F}}(\Lambda^2, T).$$

Proof: It suffices to show that both groups in question are annihilated by 2 and 3. Since the composition of the obvious transformations

$$\begin{array}{ccccc} \Lambda^2 & \longrightarrow & \otimes^2 & \longrightarrow & \Lambda^2, \\ \Gamma & \longrightarrow & \otimes^2 & \longrightarrow & \Gamma \end{array}$$

given by the multiplication by 2 it follows from lemma (8.5.4) that 2 indeed annihilates both groups. Furthermore $Ext^*_{\mathcal{F}}(\Gamma, Id^{\otimes 3} \otimes M) = 0 = Ext^*_{\mathcal{F}}(\Lambda^2, Id^{\otimes 3} \otimes M)$ by lemma (8.5.1). So the lemma follows from the fact that the composition of the obvious transformations

$$L_3(A,1)_3 \otimes M \longrightarrow A^{\otimes 3} \otimes M \longrightarrow L_3(A,1)_3 \otimes M$$

is the multiplication by 3. □

8.5.6. LEMMA:

$$Ext^*_{\mathcal{F}}(\Lambda^2 \otimes \mathbb{Z}/2, T) = 0.$$

Proof: Follows from lemma (8.5.5) and from the short exact sequence

$$0 \to \Lambda^2 \to \Lambda^2 \to \Lambda^2 \otimes \mathbb{Z}/2 \to 0.$$

□

8.5.7. LEMMA: *$Ext^1_{\mathcal{F}}(\Lambda^2 \otimes Id, T) = 0$ and $Hom_{\mathcal{F}}(\Lambda^2 \otimes Id, T) \simeq M$.*

Proof: Thanks to the proof of lemma (8.5.2) we know that $\underline{Ext}^*(Id, T) = B_1^* \oplus B_2^* \otimes Id^{\otimes 2}$, where B_1 and B_2 are constant functors. Recall that $B_2^* = Ext^*_{\mathcal{F}}(Id, M \otimes Id)$, which was computed in [FP]. Here we only need to know that

$$B_2^0 = M, \; B_2^1 = 0 = B_2^2.$$

Putting $S = \Lambda^2$, $R = Id$ and $Q = T$ in the spectral sequence (8.5.3) one obtains the isomorphism

$$Ext^p_{\mathcal{F}}(\Lambda^2, Id^{\otimes 2} \otimes M) \cong Ext^p_{\mathcal{F}}(\Lambda^2 \otimes Id, T), \tag{8.5.8}$$

provided $0 \leq p \leq 2$. Clearly Λ^2 and $Id^{\otimes 2} \otimes M$ are quadratic functors. According to the main result of [BaQ] one has an equivalence between the category of quadratic functors $\mathbf{ab} \to \mathbf{Ab}$ and the category $\mathbf{Quad}$ of quadratic $\mathbb{Z}$-modules. Moreover

$$\begin{aligned} \mathbb{Z}^{\wedge} &= (0 \longrightarrow \mathbb{Z} \longrightarrow 0), \\ \mathbb{Z}^{\otimes 2} \otimes M &= (M \xrightarrow{(i)} M \oplus M \xrightarrow{(1,1)} M) \end{aligned}$$

correspond respectively to Λ^2 and $Id^{\otimes 2} \otimes M$. The trivial calculations show that

$$Hom_{\mathbf{Quad}}(\mathbb{Z}^{\wedge}, \mathbb{Z}^{\otimes 2} \otimes M) = 0$$

and

$$Ext^1_{\mathbf{Quad}}(\mathbb{Z}^{\wedge}, \mathbb{Z}^{\otimes 2} \otimes M) = 0.$$

Clearly

$$Ext^i_{\mathcal{F}}(\Lambda^2, Id^{\otimes 2} \otimes M) = Ext^i_{\mathbf{Quad}}(\mathbb{Z}^{\wedge}, \mathbb{Z}^{\otimes 2} \otimes M), \; i = 0, 1,$$

and the result follows by (8.5.8). □

REFERENCES

[BAH] Baues, H.J.: *Algebraic Homotopy*, Cambridge Studies in Advanced Math. 15, Cambridge University Press (1987) 460 pages

[BaP] Baues, H.J.: and Pirashvili, T.: A universal coefficient theorem for quadratic functors, *J. Pure and Appl. Algebra* 148 (2000) 151–179

[BaQ] Baues, H.J.: Quadratic functors and metastable homotopy, *J. Pure and Appl. Algebra*, 94 (1994) 49–107

[BCC] Baues, H.J.: *Commutator Calculus and Groups of Homotopy Classes*, London Math. Soc. Lecture Notes 50, Cambridge University Press (1981) 160 pages

[BCH] Baues, H.J.: *Combinatorial Homotopy and 4-Dimensional Complexes*, DeGruyter Verlag Berlin (1991) 380 pages

[BCU] Baues, H.J.: On the cohomology of categories, universal Toda brackets and Homotopy pairs, *K-Theory* 11 (1997) 259–285

[BD] Baues, H.J. and Dreckmann, W.: The cohomology of homotopy categories and the general linear group, *K-Theory* 3, (1989) 307–338

[Be] Betley, S.: Calculations in THH-theory, *J. Algebra* 180 (1996) 445–458

[BEHP] Baues, H.J.: Relationen für primäre Homotopieoperationen und eine verallgemeinerte EHP-Sequenz, *Ann. scien. Ec. Norm. Sup.* 8 (1975) 509–533

[BeP] Betley, S. and Pirashvili, T.: Stable K-theory as a derived functor, *J. Pure and Appl. Algebra* 96 (1994) 245–258

[BG] Baues, H.J. and Goerss, P.: A homotopy operation spectral sequence for the computation of homotopy groups, *Topology* 39 (2000) 161–192

[BHH] Baues, H.J.: *Homotopy Type and Homology*, Oxford Math. Monograph, Oxford University Press (1996) 496 pages

[BJA] Baues, H.J. and Jibladze, M.: Classification of abelian track categories, Preprint MPI (2000)

[BJR] Baues, H.J. ans Jibladze, M.: Representability of tracks, Preprint MPI (2000)

[BNA] Baues, H.J.: Non-abelian extensions and homotopies, *K-Theory* 10 (1996) 107–133

[BW] Baues, H.J.:On homotopy clssification problems of J.H.C. Whitehead, *Algebraic Topology: Göttingen*, Lect. Notes in Math. 1172, Springer Verlag Berlin (1985)

[CH] Cochran, T. and Habegger, N.: On the homotopy theory of simply connected four manifolds, *Topology* 29 (1990) 419–440

[F] Freedman, M.H.: The topology of four dimensional manifolds, J. Differential Geometry 17 (1982) 357–453

[FP] Franjou, V. and Pirashvili, T.: On Mac Lane cohomology for the ring of integers, Topology 37 (1998) 109–114

[HKN] Hirzebruch, F. and Neumann, W.D. and Koh, S.S.: *Differentiable manifolds and quadratic forms.* Lect. Notes in pure and applied math. Marcel Dekkar, Inc. New York (1971)

[HS] Hilton, P.J. and Stammbach, U.: A course in homological algebra, Springer Verlag (1971) 338 pages

[JP] Jibladze, M. and Pirashvili, T.: Cohomolgy of algebraic theories, *J. Algebra* 137 (1991) 253–296

[K] Kreck, M.: Isotopy classes of diffeomorphisms of $(k-1)$-connected almost-parallelizable $2k$-manifolds, Lect. Notes in Math. 763 (1979) 643–650 *Algebraic Toplogy Aarhus* 1078

[KH] Kreck, M.: H-cobordism between 1-connected 4-manifolds, Preprint 2000

[KS] Kahn, P.: Self-equivalences of $(n-1)$-connected $2n$-manifolds, *Math. Annln* 180 (1969) 26–47

[M] Mandlebaum, R.: Four-dimensional topology: an introduction, *Bull Am. Math. Soc.* 2(1) (1980) 1–160

[Mac] MacLane, S.: *Categories for the working mathematician*, Graduate Texts in Math. 5, Springer Verlag, New York (1998)

[MI] Milnor, J.: On simply-connected 4-manifolds, *Symposia International Topologia Alg.* (1958)

[MH] Milnor, J. and Husemoller, D.: *Symmetric Bilinar Forms*, Springer Verlag (1973)

[Q] Quinn, F.: Isotopy of 4-manifolds, *J. Differential Geometry* 24 (1986) 343-372

[R] Rochlin, V.A.: New results in the theory of 4-dimensional manifolds, *Doklady Akad. Nauk. SSSR (N.S.)* 84 (1952) 221–224

[U] Unsöld, H.M.: A_n^4-polyhedra with free homology, *Manuscripta math.* 65 (1989) 123-145

[Ul] Ulmer, F.: *Kan extensions, cotriple and André (co)homology*, Lect. Notes in Math., Springer Verlag Berlin 92 (1969) 278–308

[W] Whitehead, J.H.C.: A certain exact sequence, *Ann. of Math.* 52 (1950) 52–110

[W4] Whitehead, J.H.C.: On simply connected 4-dimensional polyhedra, *Comment. math. Helv.* 22 (1949) 48–92

[Wa] Wall, C.T.C.: Surgery on compact manifolds, London Math. Soc. Monographs 1 Academic Press New York (1970)

[WH] Whitehead, G.W.: *Homotopy Theory*, Graduate Tesxts in Math. Springer Verlag Berlin 61 (1978)

INDEX

Printed in the United States
by Baker & Taylor Publisher Services